AF368919

Remote Sensing of Snow and
Its Applications

Remote Sensing of Snow and Its Applications

Editors

Ali Nadir Arslan
Zuhal Akyurek

MDPI • Basel • Beijing • Wuhan • Barcelona • Belgrade • Manchester • Tokyo • Cluj • Tianjin

Editors
Ali Nadir Arslan   Zuhal Akyurek
Finnish Meteorological Institute   Middle East Technical University
Finland   Turkey

Editorial Office
MDPI
St. Alban-Anlage 66
4052 Basel, Switzerland

This is a reprint of articles from the Special Issue published online in the open access journal *Geosciences* (ISSN 2076-3263) (available at: https://www.mdpi.com/journal/geosciences/special_issues/Remot_Sensing_Snow_Applications).

For citation purposes, cite each article independently as indicated on the article page online and as indicated below:

LastName, A.A.; LastName, B.B.; LastName, C.C. Article Title. *Journal Name* **Year**, *Volume Number*, Page Range.

ISBN 978-3-0365-0070-6 (Hbk)
ISBN 978-3-0365-0071-3 (PDF)

Contents

About the Editors

Ali Nadir Arslan received his D.Sc. (Tech.) degree from Helsinki University of Technology, Espoo, Finland, in 2006. He worked in Nokia Corporation between 1999 and 2009, as a principal scientist. He is currently a Senior Scientist at the Finnish Meteorological Institute (FMI) and has been employed since 2009. His research interests are remote sensing methods and applications for cryosphere, microwaves, electromagnetics theory and computational modeling, and EMC; signal and power integrity. Dr. Arslan has directed several research projects including for the European Union and some university collaborations. He has authored over 50 scientific papers and reports and he has 3 patents.

Zuhal Akyurek is a full-time professor in the Civil Engineering Department at the Middle East Technical University. She received her PhD from the Middle East Technical University, her post-doc from the Geography Department of Bristol University, and was a Visiting Scientist at NASA in 2008. She worked for the Scientific and Technological Research Council of Turkey as a researcher. She does research in hydrology, rainfall-runoff modeling, optical and microwave remote sensing of snow, and flood modeling. She gives courses in Water Resources Engineering, Hydrology, Geographic Information Systems, and Urban Hydrology and Hydraulics. Prof. Akyurek is a member of the National Hydrology Commission. She has taken part in many EU, National Scientific Research Council funded projects. She has authored more than 50 publications in peer-reviewed journals and more than 100 presentations in conferences, workshops, and seminars.

Editorial

Special Issue on Remote Sensing of Snow and Its Applications

Ali Nadir Arslan [1],* and Zuhal Akyürek [2],*

[1] Finnish Meteorological Institute (FMI), Erik Palménin aukio 1, FI-00560 Helsinki, Finland
[2] Department of Civil Engineering, Middle East Technical University, 06800 Ankara, Turkey
* Correspondence: ali.nadir.arslan@fmi.fi (A.N.A.); zakyurek@metu.edu.tr (Z.A.)

Received: 18 June 2019; Accepted: 20 June 2019; Published: 24 June 2019

Abstract: Snow cover is an essential climate variable directly affecting the Earth's energy balance. Snow cover has a number of important physical properties that exert an influence on global and regional energy, water, and carbon cycles. Remote sensing provides a good understanding of snow cover and enable snow cover information to be assimilated into hydrological, land surface, meteorological, and climate models for predicting snowmelt runoff, snow water resources, and to warn about snow-related natural hazards. The main objectives of this Special Issue, "Remote Sensing of Snow and Its Applications" in Geosciences are to present a wide range of topics such as (1) remote sensing techniques and methods for snow, (2) modeling, retrieval algorithms, and in-situ measurements of snow parameters, (3) multi-source and multi-sensor remote sensing of snow, (4) remote sensing and model integrated approaches of snow, and (5) applications where remotely sensed snow information is used for weather forecasting, flooding, avalanche, water management, traffic, health and sport, agriculture and forestry, climate scenarios, etc. It is very important to understand (a) differences and similarities, (b) representativeness and applicability, (c) accuracy and sources of error in measuring of snow both in-situ and remote sensing and assimilating snow into hydrological, land surface, meteorological, and climate models. This Special Issue contains nine articles and covers some of the topics we listed above.

Keywords: remote sensing; snow parameters; spatial and temporal variability of snow; snow hydrology; integration of remote sensing with models (hydrological; land surface; meteorological and climate)

---

1. Introduction

Snow cover is an essential climate variable directly affecting the Earth's energy balance. Snow cover has a number of important physical properties that exert an influence on global and regional energy, water, and carbon cycles. Surface temperature is highly dependent on the presence or absence of snow cover, and temperature trends have been shown to be related to changes in snow cover [1,2]. Its quantification in a changing climate is thus important for various environmental and economic impact assessments. Identification of snowmelt processes could significantly support water management, flood prediction, and prevention. Remote sensing provides a good understanding of snow cover and enables snow cover information to be assimilated into hydrological, land surface, meteorological, and climate models for predicting snowmelt runoff, snow water resources, to warn about snow-related natural hazards, and for short and long term weather forecasting. This Special Issue invited and encouraged the submission of studies covering all instrumentation/sensors and methodologies/models/algorithms in remote sensing of snow cover parameters (snow extent, snow depth, snow wetness, snow density, snow water equivalent, etc.) and applications where remotely-sensed snow information are used. Our motivation for publishing this Special Issue is to combine all aspects of remote sensing of snow from data retrieval to application. This Special Issue, "Remote Sensing of Snow and Its Applications" [3], contains nine published articles. This guest editorial addresses article contributions in this Special Issue in three categories: (a) New opportunities (Copernicus Sentinels) and emerging remote sensing methods, (b) the use of snow data in modeling, and (c) the characterization of snowpack.

2. Remote Sensing of Snow and Its Applications

2.1. New Opportunities (Copernicus Sentinels) and Emerging Remote Sensing Methods

Copernicus is the European Union (EU)'s Earth Observation (EO) program which offers free and open information services based on satellite Earth Observation and in situ (non-space) data [4]. Since the launch of Sentinel-1A [5] in 2014, the European Union set in motion a process to place a constellation of almost 20 more satellites, carrying a range of technologies such as radar and multi-spectral imaging instruments, in orbit before 2030, and that was to be implemented by the European Space Agency (ESA). These new series of satellites from the Copernicus program are very important for Remote Sensing and its applications. In order to see the status on use of Copernicus Sentinels on snow in general we looked at the published papers in Web of Science during the last five years (2015–2019). We used key words, snow and remote sensing, in searching published papers in Web of Science. A total of 313 published papers were listed by 17:00 (CET), June 7, 2019. A total of 194 out of 313 published papers were found to be related to remote sensing of snow after looking at abstracts and full papers. The papers related to remote sensing of snow using Advanced Very High Resolution Radiometer (AVHRR), Landsat, Moderate Resolution Imaging Spectroradiometer (MODIS), Passive Microwave (PMW) like Advanced Microwave Scanning Radiometer-Earth Observing System (AMSR-E), Sentinel-1 and -2, and emerging technologies are given in Table 1. Of course our objective here is not to give exact statistics but rather to present a general picture. Although there may be errors in the exact numbers of the results, we believe they are good enough to be presented here.

Table 1. Number of published papers on remote sensing of snow between 2015–2019 in Web of Science.

AVHRR	Landsat	MODIS	PMV (Mostly AMSR-E)	Sentinel-1	Sentinel-2	Emerging Technologies (UAS-Drone, GNSS, GNSS-R, GPS-IR, Webcam-Camera)
6	21	97	38	9	5	18

The MODIS and PMW were the most used ones in remote sensing snow studies according to our quick analysis. We also see that Copernicus Sentinels and emerging technologies are taking their place in remote sensing snow studies. During a PhD course entitled, "Remote Sensing and It's applications in Cryospheric sciences" given by Dr. Ali Nadir Arslan at the Department of Civil and Environmental Engineering, Politecnico di Milano, PhD students under supervision of Professor Dr Carlo de Michele conducted a user experience study on Copernicus Data Information and Access Systems (DIAS) for Earth Observation (EO) newcomers, where five online platforms allowed users to discover, manipulate, process, and download Copernicus data and information [6]. The purpose of the study was to understand possible ways to lower the data access barriers for this category of users. The following criteria are identified from the study given in Table 2.

Table 2. Criterias for lowering barriers of EO newcomers on data access systems.

User Interface	Database Repository	Database Acquisitions & Services
Simplicity & Clarity	Dataset Services & Descriptions	Advanced Products/Tools/Services
Demonstarte Core Benefits Effectively	Other Dataset than Copernicus	Cloud Services
User Guides & Tutorials	Search Criteria by Region	Ease of Downloading
Help Desk & FAQ	Search Criteria by Products	Example Analysis
User Update	Search Criteria by End-use Application	Customized/Direct Pricing
No Prerequisite Knowledge	Visualization by Timeline	Open Source Software
Mobile Compatibility	Global/Regional Visualization	Monitoring & Dashboard

There has been a tradeoff between the spatial and temporal resolution of remote sensing snow mapping because of the characteristics of available optical and microwave (passive and active) sensors used for snow detection [7]. Medium spatial resolution satellite-derived snow products are good in monitoring snow dynamics, but a better spatial resolution is needed in understanding the spatial variation, especially in rough terrain. Validation of the satellite-derived snow products is also very important and critical. The in situ snow observations may not be representing the field of view of the satellite data. This mismatching problem can be solved by using remote sensing data with high spatial resolution. Piazzi et al. [8] presented the use of Sentinel-2A high resolution satellite data in validating the moderate resolution satellite-derived snow products, namely H10 and H12 supplied by the Satellite Application Facility on Support to Operational Hydrology and Water Management (H-SAF) project of the European Organization for the Exploitation of Meteorological Satellites (EUMETSAT). In their study they used the webcam imagery as ground data. Salzano et al. [9] presented the importance of ground based cameras in obtaining a long time series of snow cover. The images taken over a ten-year period were analyzed using an automated snow-not-snow detection algorithm based on Spectral Similarity. Webcam/camera monitoring system is easy to implement and very cheap in comparison to satellites. It provides a valuable data source and can be used as complementary information to different multispectral remotely sensed datasets for validation and calibration processes [9]. The usage of webcam on monitoring snow is growing [10–15] and there are already several tools [16,17] available for processing webcam/camera data.

MODIS snow products provide a good archive since 2000 and, as presented in Table 1, MODIS products have been used in many snow studies. Munkhjargal et al. [18] combined MODIS snow products with Landsat-7 and -8, and Sentinel-2A images to map snow cover and its duration in the mountainous region with 30 m resolution by applying a series of adjustments, including temporal gap-filling and conditional adjustments. In understanding climate related glacier behavior, continuous monitoring of glacier changes is needed. Heilig et al. [19] presented the use of Sentinel-1 data in monitoring the recession of wet snow area extent per season for three different glacier areas of the Rofental, Austria. They showed that surface conditions during the melt season can quasi-continuously be monitored using Sentinel-1 SAR data, which is essential for glacier runoff modeling.

2.2. The Use of Snow Data in Modeling

The in situ and satellite snow data contribute to the development of both retrieval algorithms of remote sensing and snowpack models as they are used in validation studies. Recently, data assimilation (DA) has provided an outstanding solution for improving hydrological modeling by synchronously integrating observations from in situ stations or remote sensors. Helmert et al. [20] reviewed approaches used for assimilation of snow measurements, such as remotely sensed and in situ observations into hydrological, land surface, meteorological, and climate models based on a COST HarmoSnow survey exploring the common practices on the use of snow observation data in different modeling environments. As concluded by other authors [21–23], this study highlights the need of assimilation of bias corrected snow data to get consistently improved snow and streamflow predictions. Therefore, it is essential to improve the snow observation data quality before assimilation into hydrological models, otherwise the model performance will deteriorate. Arsenault and Houser [24] presented a new approach to estimating snow depletion curves (SDC) and their application for assimilating snow cover fraction observations using an Ensemble Kalman filter (EnKF) data assimilation approach and a land surface model with a multi-layer snow physics scheme. They presented that the use of observation-based SDC (they derived the new SDC from the MODIS snow cover fraction and SNOTEL snow water equivalent (SWE) observations) showed improvement over the default model-based snow cover fraction (SCF) forecasts and snow state analysis. Appel et al. [25] used the in situ and EO information to assimilate the input and the parameters of the applied hydrological model PROMET (Processes of Mass and Energy Transfer) to calculate SWE, snowmelt onset, and river run-off in catchments as spatial layers. They used newly developed in situ snow monitoring stations based on signals of the Global Navigation Satellite System (GNSS) and Sentinel-1A and -1B EO data in interferometric wide (IW) swath mode on the snow cover extent and on information whether the snow is dry or wet. The snow monitoring stations based on signals of the GNSS is a state-of-the-art remote sensing techniques which was mentioned in [7]. It is known that for hydrologists, snow mass and volume parameters are more critical, because the water volume stored in the snowpack and subsequent snow melt runoff can be estimated. Most in situ snow measurements are still performed using traditional laborious standardized techniques: Sampling with snow tubes, digging snow pits, and manually measuring the density, temperature, hardness, and other quantities. While these techniques are very robust and straightforward, they are very expensive for larger areas or time spans, prone to human errors and biases, and do not provide all requested quantities or provide only qualitative information of snow parameters. Obtaining continuous snow water equivalent at the field is still missing and challenging. Cosmic ray sensors [26] and snowpack analyzers [27] are two new techniques that can be used in the field. Both of the instruments need more validation studies to be considered as robust in situ SWE measurement techniques.

2.3. Characterization of Snowpack

It is a challenging problem of bridging information from micro-structural scales of the snowpack up to the grid resolution in models. In-situ ground-truth snow observations are necessary for developing and validating remote sensing products. The advances in the modeling of the snow-electromagnetic interaction and in the observational capabilities of the satellite-based sensors have pushed the development of new in situ instrumentation, which are able to provide suitable reference and ground-truth data for the validation of snow satellite products and of earth system models [28]. Monitoring of snow extent and SWE requires solid knowledge of the physical properties of snow, high-quality instrumentation, and refined methods for calibration and interpretation of snow observations. Leppanen et al. [29] presented an empirical linear relationship, on taiga snow, between specific surface area (SSA) and reflectance observations of recently developed hand-held QualitySpec Trek (QST) instrument. The microstructure of snow is an important parameter for the modeling of microwave emission and optical reflectance, and it is therefore also important for remote sensing applications. The SSA is an important snow parameter for the modeling of microwave emission and optical reflectance, and is therefore also important for remote sensing applications [29]. Sanow et al. [30] presented terrestrial laser scanner- (TLS) (resolution of +/−5 mm) derived surface geometry and vertical wind profile measurements to compare concurrent aerodynamic roughness length estimates for changing snow surface features of shallow snowpack. The roughness of a snow surface is an important control on air-snow heat transfer and changes in the snow surface can have substantial effects on the energy balance at this interface [30].

3. Summary

Monitoring the snow cover and its components at meso-, regional to global scale is important in order to support weather, hydrological, and climate science, as well as in the monitoring of natural hazards, and the decision-making and formulation of environmental policy. This capability will provide knowledge-based information on potential impacts to society, economy, and safety (e.g., hydro-power, water availability, transportation, tourism, flooding, avalanches, etc.). Snow is a complex media which is why all aspects such as characterization, sensing, and modeling are important to understand. Our aim was to combine these three aspects together as we believe this will be useful for all disciplines dealing with some part of snow science. Although this is a very small effort, we hope that this will be useful for the scientific community.

Author Contributions: Writing, review, and editing A.N.A and Z.A.

Acknowledgments: We thank all the authors of the published papers in this special issue and we would also like to give a special thanks to the Geosciences and its team for having this special issue and for all their support and patience.

References

1. Lemke, P.; Ren, J.; Alley, R.B.; Allison, I.; Carrasco, J.; Flato, G.; Fujii, Y.; Kaser, G.; Mote, P.; Thomas, R.H.; et al. Observations: Changes in Snow, Ice and Frozen Ground. In *Climate Change 2007; The Physical Science Basis. Contribution of Working Group I to the Fourth Assessment Report of the Intergovernmental Panel on Climate Change*; Solomon, S., Qin, D., Manning, M., Chen, Z., Marquis, M., Averyt, K., Tignor, M., Miller, H.L., Eds.; Cambridge University Press: Cambridge, UK; New York, NY, USA, 2007.

2. Climate Change: Global Temperature. Available online: https://www.climate.gov/news-features/understanding-climate/climate-change-global-temperature (accessed on 13 June 2019).

3. Special Issue Remote Sensing of Snow and Its Applications. Available online: https://www.mdpi.com/journal/geosciences/special_issues/Remot_Sensing_Snow_Applications (accessed on 13 June 2019).

4. Copernicus. Available online: https://www.copernicus.eu/ (accessed on 1 June 2019).

5. Sentinels. Available online: https://www.esa.int/Our_Activities/Observing_the_Earth/Copernicus (accessed on 1 June 2019).

6. Copernicus DIAS. Available online: https://www.copernicus.eu/en/access-data/dias (accessed on 12 June 2019).

7. Dong, C. Remote Sensing, Hydrological Modeling and In Situ Observations in Snow Cover Research: A review. *J. Hydrol.* **2018**, *561*, 573–583. [CrossRef]

8. Piazzi, G.; Tanis, C.M.; Kuter, S.; Simsek, B.; Puca, S.; Toniazzo, A.; Takala, M.; Akyürek, Z.; Gabellani, S.; Arslan, A.N. Cross-Country Assessment of H-SAF Snow Products by Sentinel-2 Imagery Validated against In-Situ Observations and Webcam Photography. *Geosciences* **2019**, *9*, 129. [CrossRef]

9. Salzano, R.; Salvatori, R.; Valt, M.; Giuliani, G.; Chatenoux, B.; Ioppi, L. Automated Classification of Terrestrial Images: The Contribution to the Remote Sensing of Snow Cover. *Geosciences* **2019**, *9*, 97. [CrossRef]

10. Salvatori, R.; Plini, P.; Giusto, M.; Valt, M.; Salzano, R.; Montagnoli, M.; Cagnati, A.; Crepaz, G.; Sigismondi, D. Snow cover monitoring with images from digital camera systems. *Ital. J. Remote Sens.* **2011**, *43*, 137–145. [CrossRef]

11. Parajka, J.; Haas, P.; Kirnbauer, R.; Jansa, J.R.; Blöschl, G. Potential of time-lapse photography of snow for hydrological purposes at the small catchment scale. *Hydrol. Process.* **2012**, *26*, 3327–3337. [CrossRef]

12. Bernard, E.; Friedt, J.M.; Tolle, F.; Griselin, M.; Martin, G.; Laffly, D.; Marlin, C. Monitoring seasonal snow dynamics using ground based high resolution photography (Austre Lovénbreen, Svalbard, 79 N). *ISPRS J. Photogramm. Remote Sens.* **2013**, *75*, 92–100. [CrossRef]

13. Garvelmann, J.; Pohl, S.; Weiler, M. From observation to the quantification of snow processes with a time-lapse camera network. *Hydrol. Earth Syst. Sci.* **2013**, *17*, 1415–1429. [CrossRef]

14. Fedorov, R.; Camerada, A.; Fraternali, P.; Tagliasacchi, M. Estimating snow cover from publicly available images. *IEEE Trans. Multimed.* **2016**, *18*, 1187–1200. [CrossRef]

15. Arslan, A.N.; Tanis, C.M.; Metsämäki, S.; Aurela, M.; Böttcher, K.; Linkosalmi, M.; Peltoniemi, M. Automated Webcam Monitoring of Fractional Snow Cover in Northern Boreal Conditions. *Geosciences* **2017**, *7*, 55. [CrossRef]

16. Härer, S.; Bernhardt, M.; Corripio, J.G.; Schulz, K. PRACTISE—Photo Rectification and ClassificaTIon SoftwarE (V.1.0). *Geosci. Model Dev.* **2013**, *6*, 837–848. [CrossRef]

17. Tanis, C.M.; Peltoniemi, M.; Linkosalmi, M.; Aurela, M.; Böttcher, K.; Manninen, T.; Arslan, A.N. A system for Acquisition, Processing and Visualization of Image Time Series from Multiple Camera Networks. *Data* **2018**, *3*, 23. [CrossRef]

18. Munkhjargal, M.; Groos, S.; Pan, C.G.; Yadamsuren, G.; Yamkin, J.; Menzel, L. Multi-Source Based Spatio-Temporal Distribution of Snow in a Semi-Arid Headwater Catchment of Northern Mongolia. *Geosciences* **2019**, *9*, 53. [CrossRef]

19. Heilig, A.; Wendleder, A.; Schmitt, A.; Mayer, C. Discriminating Wet Snow and Firn for Alpine Glaciers Using Sentinel-1 Data: A Case Study at Rofental, Austria. *Geosciences* **2019**, *9*, 69. [CrossRef]

20. Helmert, J.; Şensoy Şorman, A.; Alvarado Montero, R.; De Michele, C.; De Rosnay, P.; Dumont, M.; Finger, D.C.; Lange, M.; Picard, G.; Potopová, V.; et al. Review of Snow Data Assimilation Methods for Hydrological, Land Surface, Meteorological and Climate Models: Results from a COST HarmoSnow Survey. *Geosciences* **2018**, *8*, 489. [CrossRef]

21. Slater, A.G.; Clark, M. Snow Data Assimilation via an Ensemble Kalman Filter. *J. Hydrometeorol.* **2006**, *7*, 478–493. [CrossRef]

22. Thirel, G.; Salamon, P.; Burek, P.; Kalas, M. Assimilation of MODIS Snow Cover Area Data in a Distributed Hydrological Model Using the Particle Filter. *Remote Sens.* **2013**, *5*, 5825–5850. [CrossRef]

23. Liu, Y.; Peters-Lidard, C.D.; Kumar, S.; Foster, J.L.; Shaw, M.; Tian, Y.; Fall, G.M. Assimilating satellite-based snow depth and snow cover products for improving snow predictions in Alaska. *Adv. Water Resour.* **2013**, *54*, 208–227. [CrossRef]

24. Arsenault, K.R.; Houser, P.R. Generating Observation-Based Snow Depletion Curves for Use in Snow Cover Data Assimilation. *Geosciences* **2018**, *8*, 484. [CrossRef]

25. Appel, F.; Koch, F.; Rösel, A.; Klug, P.; Henkel, P.; Lamm, M.; Mauser, W.; Bach, H. Advances in Snow Hydrology Using a Combined Approach of GNSS In Situ Stations, Hydrological Modelling and Earth Observation—A Case Study in Canada. *Geosciences* **2019**, *9*, 44. [CrossRef]

26. Zreda, M.; Köhli, M.; Schrön, M.; Hamann, S.; Womack, G. *Using Downward-Looking Cosmogenic Neutron Sensor to Calibrate Wide-Area Sensor and to Measure Snow Water Equivalent*; EGU General Assembly: Vienna, Austria, 2019; Volume 21.

27. Snowpack Analyzers. Available online: https://www.sommer.at/en/ (accessed on 14 June 2019).

28. Pirazzini, R.; Leppänen, L.; Picard, G.; Lopez-Moreno, J.I.; Marty, C.; Macelloni, G.; Kontu, A.; von Lerber, A.; Tanis, C.M.; Schneebeli, M.; et al. European In-Situ Snow Measurements: Practices and Purposes. *Sensors* **2018**, *18*, 2016. [CrossRef]

29. Leppänen, L.; Kontu, A. Analysis of QualitySpec Trek Reflectance from Vertical Profiles of Taiga Snowpack. *Geosciences* **2018**, *8*, 404. [CrossRef]

30. Sanow, J.E.; Fassnacht, S.R.; Kamin, D.J.; Sexstone, G.A.; Bauerle, W.L.; Oprea, I. Geometric Versus Anemometric Surface Roughness for a Shallow Accumulating Snowpack. *Geosciences* **2018**, *8*, 463. [CrossRef]

geosciences

Article

Cross-Country Assessment of H-SAF Snow Products by Sentinel-2 Imagery Validated against In-Situ Observations and Webcam Photography

Gaia Piazzi [1],[*],[†], Cemal Melih Tanis [2], Semih Kuter [3], Burak Simsek [2], Silvia Puca [4], Alexander Toniazzo [4], Matias Takala [2], Zuhal Akyürek [5], Simone Gabellani [1] and Ali Nadir Arslan [2]

[1] CIMA Research Foundation, 17100 Savona, Italy; simone.gabellani@cimafoundation.org
[2] Finnish Meteorological Institute (FMI), 00560 Helsinki, Finland; Cemal.Melih.Tanis@fmi.fi (C.M.T.); burak.simsek@fmi.fi (B.S.); Matias.Takala@fmi.fi (M.T.); ali.nadir.arslan@fmi.fi (A.N.A.)
[3] Department of Forest Engineering, Faculty of Forestry, Çankırı Karatekin University, Çankırı 18200, Turkey; semihkuter@yahoo.com
[4] National Civil Protection Department, 00189 Rome, Italy; Silvia.Puca@protezionecivile.it (S.P.); Alexander.Toniazzo@protezionecivile.it (A.T.)
[5] Department of Civil Engineering, Middle East Technical University, Ankara 06800, Turkey; zakyurek@metu.edu.tr
* Correspondence: gaia.piazzi@irstea.fr
† Current address: IRSTEA, Hydrology Research Group, UR HYCAR, 92761 Antony, France.

Received: 2 December 2018; Accepted: 11 March 2019; Published: 15 March 2019

Abstract: Information on snow properties is of critical relevance for a wide range of scientific studies and operational applications, mainly for hydrological purposes. However, the ground-based monitoring of snow dynamics is a challenging task, especially over complex topography and under harsh environmental conditions. Remote sensing is a powerful resource providing snow observations at a large scale. This study addresses the potential of using Sentinel-2 high-resolution imagery to assess moderate-resolution snow products, namely H10—Snow detection (SN-OBS-1) and H12—Effective snow cover (SN-OBS-3) supplied by the Satellite Application Facility on Support to Operational Hydrology and Water Management (H-SAF) project of the European Organisation for the Exploitation of Meteorological Satellites (EUMETSAT). With the aim of investigating the reliability of reference data, the consistency of Sentinel-2 observations is evaluated against both in-situ snow measurements and webcam digital imagery. The study area encompasses three different regions, located in Finland, the Italian Alps and Turkey, to comprehensively analyze the selected satellite products over both mountainous and flat areas having different snow seasonality. The results over the winter seasons 2016/17 and 2017/18 show a satisfying agreement between Sentinel-2 data and ground-based observations, both in terms of snow extent and fractional snow cover. H-SAF products prove to be consistent with the high-resolution imagery, especially over flat areas. Indeed, while vegetation only slightly affects the detection of snow cover, the complex topography more strongly impacts product performances.

Keywords: snow cover; fractional snow cover; Sentinel-2; H-SAF; webcam photography

---

1. Introduction

The knowledge of the extent and location of snow cover is of key importance to enhance the understanding of the present and future climate, hydrological cycle, and ecological dynamics, at both local and global scales [1,2]. Indeed, snow-dominated regions serve as an active reservoir for water supply during the melting period [3,4], and the seasonal presence of snow cover significantly modulates a surface energy balance because of its high albedo and thermal properties [5]. Therefore, information on the spatial and temporal distribution of snow cover is critical for several research purposes and operational applications [6]. However, the monitoring of a snow-covered area is generally hindered by the complex interactions among site-dependent factors, especially in mountainous and forested regions. Meteorological forcings (i.e., precipitation regime, average air temperature, solar radiation) [7–9] and local topography (i.e., elevation, slope orientation and mean aspect) [10–12] are the most explanatory variables affecting spatial and temporal variability and persistence of snow cover. The definition of the topographic control on snow distribution is made challenging by the presence of vegetation, which intercepts snowfall and impacts the intensity of meteorological forcings [13–15]. Furthermore, wind-induced erosion and deposition phenomena are the main control factors driving the snow's spatial redistribution [16,17]. In-situ automatic measurements provide continuous and direct observational data allowing the retrieval of a temporal evolution of snow cover. However, they are site-dependent, generally subjected to distortions (e.g., wind action, vegetation interactions), and they do not succeed in catching the spatial variability of snowpack due to the heterogeneity of both climate and terrain with respect to the network density [18,19]. The collection of in-situ measurements at a large scale necessarily faces a general widespread lack of instrumental records, especially for steep slopes and remote high-elevation areas, where harsh environmental conditions usually entail a high operating cost [20].

Among in-situ gauges, the time-lapse camera is renowned for being a cost-effective device to monitor many environmental variables for scientific purposes [21–25]. Several webcam networks are currently operational worldwide, such as the European phenology camera network (EUROPhen) [26], the PhenoCam Network [27], and MONIMET camera network [24]. Recently, a growing interest aims at using webcam photography to detect snow cover from digital images to monitor its variability in space and time, even though the use of these observations is restricted to limited spatial scales [12,28–34].

Remote sensing represents a suited and powerful tool to monitor snow properties at larger scales and to overcome the gradual decrease of the representativeness of the gauging network with the increasing altitude. Under specific conditions (e.g., day-time, absence of cloudiness) [35], the snow cover detection is relatively straightforward through satellite-based optical observations, because of the high albedo of snow with respect to most land surfaces and the higher near-infrared reflectance of most clouds compared to snow-covered surfaces [36,37]. As well as cloud cover, the vegetation can obstruct visible and infrared information about snow, especially where forest canopy protruding above the snowpack reduces the surface albedo [38] and partially or completely shades the underlying surface [39,40]. Nevertheless, since satellite-based data are indirect measurements of snow-related quantities, they require a quantitative understanding of their accuracy, mainly depending on the uncertainty in retrieval algorithms [37,41]. Therefore, the comprehensive validation of satellite snow products is of key importance to properly assess and quantify their reliability, to identify possible errors and to provide input for further improvements. Indeed, the availability of information on the quality control of remotely-sensed data is critically needed by the scientific community, as one of the main key criteria for the selection of the most proper dataset to be effectively used, according to the final purpose.

Numerous studies have addressed the validation of satellite snow products at local and global scale by assessing the accuracy of remotely-sensed observations against ground-based data, which is one of the most widely used validation procedures [42–54]. Lacking any available in-situ reference data, a common approach relies on a cross-sensor comparison among different satellite-derived snow products by assuming one of the analyzed datasets as the reference truth [55–58]. This approach is even necessary when assessing the accuracy of satellite-derived products of fractional snow cover (FSC) requiring spatially distributed observations of reference [33]. Even though currently there is no agreed-upon methodology to perform a cross-sensor comparison, the most commonly used approach assumes the high-resolution satellite imagery as the reference effective dataset to assess moderate-resolution remotely-sensed observations, since it is supposed to provide the most reliable information on the actual snow cover [59]. Nowadays, the Sentinel-2 (S-2) mission of European Copernicus Earth Observation program provides high-resolution multispectral imagery with an operational short revisiting time (~5 days) and free, global and systematic availability. Because of its meaningful payload, several studies have already experienced the potentialities of S-2 data in different fields of application [60–65].

This study aims to investigate the potential use of S-2 data to assess the reliability of moderate-resolution products of snow extent and FSC, that are the snow-related quantities most commonly used as input for hydrologic, meteorological and climate modelling [2]. Indeed, H10—Snow detection (SN-OBS-1) and H12—Effective snow cover (SN-OBS-3) supplied by the EUMETSAT's H-SAF project are compared against S-2 imagery. The interest in H-SAF snow products is focused on investigating the potential of these datasets and their suitability for hydrological purposes [56]. Over past decades, operational H-SAF snow products have been continuously validated against ground-based snow measurements [53,56]. Even though the high-resolution imagery can be reasonably used to establish reliable ground truth, a finer spatial resolution does not necessarily entail a higher accuracy of the satellite product, since its accuracy strongly depends on the retrieval algorithm used to derive snow-related information. Furthermore, no existing study supplies detailed information on the accuracy of S-2 imagery in detecting snow. For these reasons, the study has therefore the dual objective of validating this high-resolution dataset against in-situ snow measurements and webcam photography in order to properly assess its consistency and to guarantee the reliability of the comparison analysis. Therefore, before addressing the cross-sensor comparison of snow satellite products, a comprehensive validation of S-2 data is performed against ground-based data. With the aim of testing and assessing the satellite snow products under different climatological and topographic conditions, three study areas located in Finland, the Italian Alps, and Turkey are analyzed.

After introducing the context of this study, its motivation and research purposes, the article consists of four main sections. Section 2 is focused on data collection through a comprehensive description of the analyzed remotely-sensed (i.e., S-2, H-SAF H10 and H12) and ground-based (i.e., snow depth measurements and webcam imagery) datasets. The selected case studies in Finland, Italy and Turkey are presented and characterized. In Section 3, the methodology is extensively explained into the details of the retrieval algorithms for the generation of satellite products. The procedures implemented within both the validation of S-2 imagery against in-situ data, and the comparison between S-2-based and moderate-resolution snow products are widely described and their main assumptions are discussed. Results are reported and assessed through several evaluation metrics in Section 4. Lastly, conclusions are outlined in Section 5.

2. Materials

2.1. Satellite Datasets

2.1.1. Sentinel-2 Imagery

S-2 is a polar-orbiting, multispectral high-resolution imaging mission of the European Space Agency (ESA) for land, ocean and atmospheric monitoring. With the aim of fulfilling revisit and coverage requirements and providing robust datasets, the constellation consists of two identical satellites, Sentinel-2A (S-2A) and Sentinel-2B (S-2B), which were launched on 23 June 2015 (operational in early 2016) and 7 March 2017, respectively. Since the twin satellites are in the same sun-synchronous orbit with a phase delay of 180°, they guarantee an effective revisit time of 5 days at the equator and 2/3 days over mid-latitudes, with a 290-km swath width. Multi-Spectral Imager (MSI) instruments provide fine spatial resolution optical images (Figure 1) having 13 bands spanning from the visible and the near infrared to the shortwave infrared, covering wavelengths from 0.4 to 2.2 µm (Table 1) [66]. Depending on the spectral band, the spatial resolution varies from 10 to 60 m. Four visible and near-infrared (VNIR) bands at 10 m for optical measurement, four NIR bands at 20 m for vegetation red-edge, two shortwave infrared (SWIR) bands at 20 m for snow, ice, and cloud discrimination, three coarse bands at 60 m in the aerosol, water vapor, and cirrus domain designated for atmospheric correction [67,68]. However, it is noteworthy that S-2 does not have a thermal band, which is of key importance for cloud detection, as cloud pixels are much colder than clear-sky pixels [69]. By December 2015, the acquisition of S-2 Level-1C (L1C) top-of-atmosphere (TOA) reflectance data is available and currently also S-2 Level-2A (L2A) bottom-of-atmosphere (BOA) reflectance data product is available to the remote sensing community worldwide.

Figure 1. Sentinel-2 RGB image—Tile T32TLR (Italian Alps), 6 December 2017.

Table 1. Spatial resolution and central wavelength of Sentinel-2A and -2B spectral bands.

Band Number	Spatial Resolution [m]	S-2A Central Wavelength [nm]	S-2B Central Wavelength [nm]
1	60	442.7	442.2
2	10	492.4	492.1
3	10	559.8	559.0
4	10	664.6	664.9
5	20	704.1	703.8
6	20	740.5	739.1
7	20	782.8	779.7
8	10	832.8	832.9
8a	20	864.7	864.0
9	60	945.1	943.2
10	60	1373.5	1376.9
11	20	1613.7	1610.4
12	20	2202.4	2185.7

2.1.2. H-SAF H10 Product

H-SAF H10 (SN-OBS-1) is a daily operational product of snow extent generated from the visible (VIS) and infrared (IR) radiometry of the Spinning Enhanced Visible and Infrared Imager (SEVIRI) instrument on board the geostationary Meteosat Second Generation (MSG) satellites. The high temporal resolution and wide aerial coverage of SEVIRI imagery make it highly suitable for snow-cover mapping, since cloud cover is continuously monitored. Indeed, the daily snow cover product is derived for a multi-temporal analysis of SEVIRI 15-min images, that are processed as new data are available to collect the largest possible number of cloud-free pixels. The sampling is performed at 3-km intervals, which degrade to ~5 km over Europe. The resulting daily map has a spatial coverage delimited between longitude 25° W–45° E and latitude 25°–75° N [56,70] and it consists of four different classes: snow, cloud, water and bare ground (Figure 2).

Figure 2. H-SAF H10 product (25° W–45° E, 25°–75° N), 5 April 2017.

2.1.3. H-SAF H12 Product

H-SAF H12 (SN-OBS-3) is a daily operational product of FSC based on the multi-channel analysis of the Advanced Very High Resolution Radiometer (AVHRR) on board National Oceanic and Atmospheric Administration (NOAA) and meteorological operational (MetOp) satellites. FSC is generated at pixel resolution by exploiting the brightness intensity, which is the convolution of the snow signal and the fraction of snow within the pixel.

The sampling is carried out at 1 km intervals over the same H-SAF area of H10 product. The thematic map includes cloud and water classes, and percentage classes of fraction snow cover ranging from 0% (i.e., snow-free condition) to 100% (i.e., full snow cover) (Figure 3).

Figure 3. H-SAF H12 product quick-look image (25° W–45° E, 25°–75° N), 5 April 2017.

2.2. Test Sites and Data Collection

With the aim of properly investigating the reliability of satellite snow products and their consistency under different topographical conditions (i.e., mountainous and flat areas) and vegetation cover, this study includes three case studies located in Finland, the Italian Alps and Turkey (Figures 4–6).

In each country, eight S-2 tiles of interest have been selected to properly ensure a sizeable sample of satellite images. The selection of S-2 tiles has targeted those providing significant datasets over the analyzed period by minimizing possible overlapping. Furthermore, when selecting S-2 tiles the location of in-situ monitoring instruments has been considered to allow the validation of S-2 imagery against ground-based data. It is noteworthy to consider that in this study both snow extent and FSC are referred to the snow cover viewable over the satellite field of view, and not at the ground level. Because of the significant impact of vegetation on satellite snow detection, ancillary information on the vegetation cover of each S-2 tile has been derived from ESA GlobCover 2009 land cover map to support the assessment of the comparison results. This GlobCover map is derived from an automated classification of the Medium Resolution Imaging Spectrometer Full Resolution (MERIS FR) time series and it consists of 22 land-cover classes at a 300 m spatial resolution. Among the classes of natural and semi-natural terrestrial vegetation, two main categories have been defined (Table 2). The first main category embraces the vegetation classes having the highest impact on snow detection (V_1) (i.e., evergreen or semi-deciduous forest), while the second one includes those having a lower impact (V_2) (i.e., deciduous forest).

The selected tiles are reported in Table 3, where the percentage values of the main vegetation categories are reported according to the GlobCover 300 m land cover map.

Table 2. Selected vegetation classes of ESA GlobCover product.

Vegetation Class	Selection of GlobCover Vegetation Classes
V_1	Closed to open (>15%) broadleaved evergreen and/or semi-deciduous forest (>5 m)
	Closed (>40%) needle-leaved evergreen forest (>5 m)
	Open (15–40%) needle-leaved deciduous or evergreen forest (>5 m)
	Closed to open (>15%) mixed broadleaved and needle-leaved forest (>5 m)
V_2	Closed (>40%) broadleaved deciduous forest (>5 m)
	Open (15–40%) broadleaved deciduous forest (>5 m)

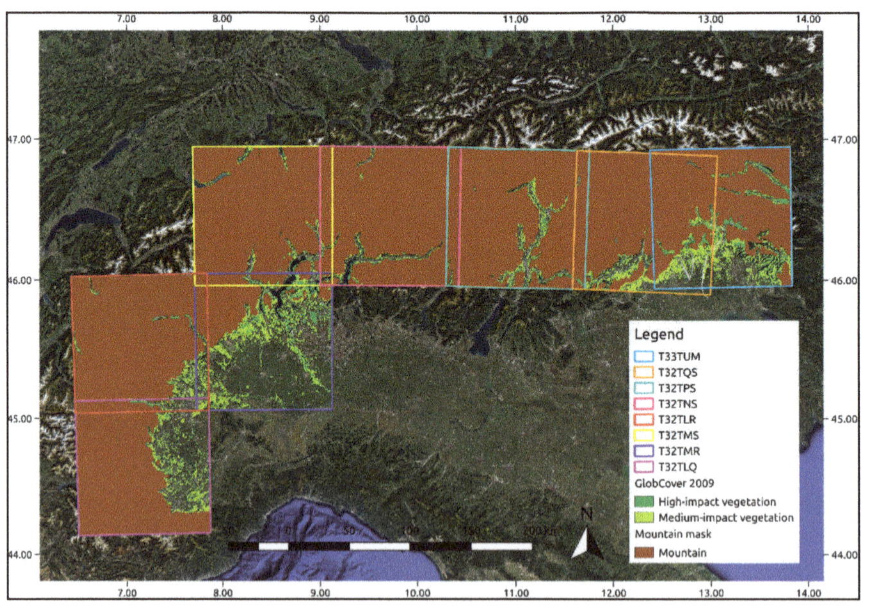

Figure 4. Selection of S-2 tiles over Italian Alps.

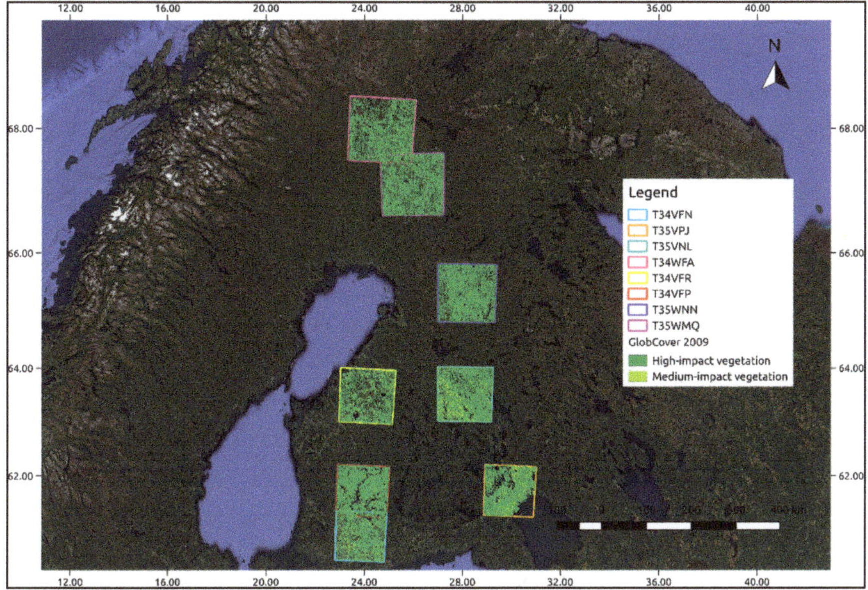

Figure 5. Selection of S-2 tiles in Finland.

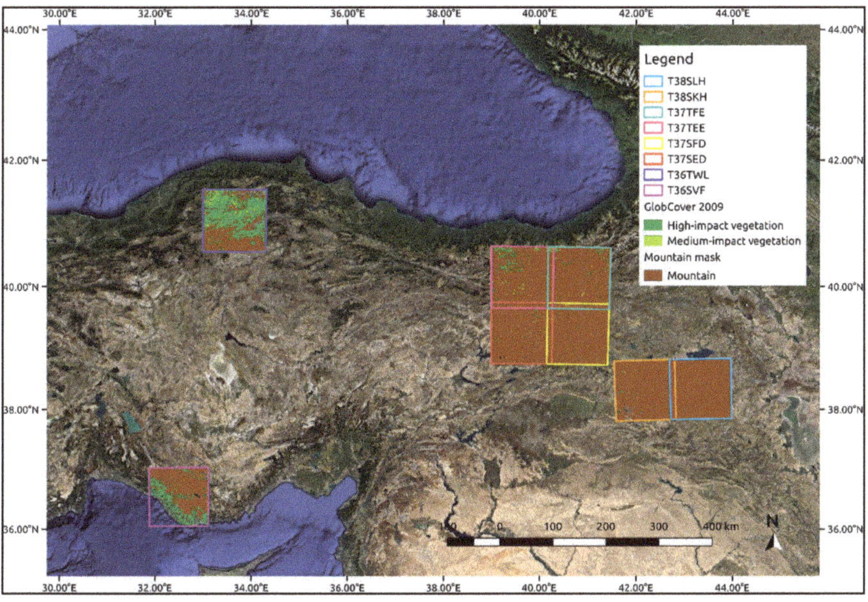

Figure 6. Selection of S-2 tiles in Turkey.

The analysis period extends throughout two winter seasons, namely 2016/2017 and 2017/2018. With the aim of properly taking account of the local climatology, the duration of the snow season has been independently set from October to May (eight months) in Finland and over the Italian Alps, and from November to April (six months) in Turkey.

Table 3. Selection of S-2 tiles at each test site and characterization of vegetation cover according to GlobCover 2009 land cover map. Two main vegetation classes having high (V_1) and medium (V_2) impact on snow detection are reported.

Selection of S-2 Tiles							
Finland	T34VFN	T34VFP	T34VFR	T34WFA	T35VNL	T35VPJ	T35WMQ	T35WNN
V_1	53%	73%	45%	57%	62%	57%	69%	73%
V_2	10%	6%	8%	2%	17%	14%	2%	2%
Italy	T32TLQ	T32TLR	T32TMR	T32TMS	T32TNS	T32TPS	T32TQS	T33TUM
V_1	10%	13%	6%	17%	20%	34%	33%	30%
V_2	15%	12%	19%	15%	12%	14%	21%	22%
Turkey	T36SVF	T36TWL	T37SED	T37SFD	T37TEE	T37TFE	T38SKH	T38SLH
V_1	17%	41%	1%	0%	5%	2%	0%	0%
V_2	0%	6%	0%	0%	2%	1%	0%	0%

Since this study is focused on assessing how satellite products succeed in detecting snow cover, cloud free scenes or scenes with minor cloud cover are primarily selected. Indeed, only S-2 images with cloud cover lower than 20% have been included in the analysis.

The resulting datasets of S-2 imagery for the analyzed case studies are reported in Table 4. It is noteworthy to consider that the effective number of S-2 images in the snow season 2017/2018 is significantly higher than in the previous one (Table 4), since S-2B data has become available in March 2017.

Table 4. Seasonal effective number of S-2 images at each test site.

Test Site	Seasonal Number of S-2 Images	
	Snow Season 2016/17	Snow Season 2017/18
Finland	60	193
Italian Alps	133	198
Turkey	37	101

Throughout the analyzed period, only one daily H10 image is missing during the first snow season and 7 images are missing in the second one. Likewise, H12 product is not available for 7 and 16 days in snow seasons 2016/17 and 2017/18, respectively.

2.3. Ground-Based Datasets

The validation of S-2 imagery relies on both ground-based dataset of snow measurements in Turkey and digital observations in Finland and over the Italian Alps.

2.3.1. In-Situ Webcam Imagery

In Finland and in Italy, in-situ webcam imagery has been used to assess the consistency of FSC maps based on S-2 data (S-2-derived FSC), which are derived by counting the number of S-2 snow pixels versus the total number of S-2 pixels over the camera FOV. Webcams have been selected according to two main criteria. The first constraint requires a sufficiently wide webcam field of view (FOV) enabling the comparison with S-2-derived FSC. Secondly, webcams providing a properly representative dataset of observations have been primarily selected. With the aim of complying with these conditions, five webcams have been selected (Table 5), only one of which located over Italian Alps, mainly due to the complex topography, which strongly limits the extent of the webcam FOV.

The four cameras selected in Northern Finland are part of the camera network deployed in the frame of the MONIMET project [24]. MONIMET monitoring network consists of 28 cameras in 14 locations in Finland. The images are free and open. Those cameras produce images at each half an hour during daytime. For the study, midday time images are used since the snow cover does not change significantly during the day. One of the cameras is located in Kenttärova looking over a large evergreen spruce forest, another one is located in Lomppolojankka, a peatland site, and the other two in Sodankylä, located in a Scots pine ecosystem and in a wetland site, respectively. The FOVs of those cameras is shown in Figure 7a–d.

The webcam located in Aosta Valley (north-western Italian Alps) is at the experimental site of Torgnon, which belongs to the Phenocam network [71]. The camera is pointed north and it looks over grassland with mountains visible at distance. Camera images are provided every hour from 10 a.m. to 4 p.m. [21]. The FOV of the camera is shown in Figure 7e.

Figure 7. Webcam field of views: (**a**) Kenttärova canopy camera, (**b**) Lompolojankka peatland camera, (**c**) Sodankylä canopy camera, (**d**) Sodankylä peatland camera, (**e**) Torgnon camera.

Table 5. Selected in-situ cameras in Finland and Italy.

Site Name	Coordinates	Camera Brand and Model	Resolution	S-2 Tile	No. of Analyzed Images
Torgnon	45.84° N, 7.57° E	Campbell CC640	0.3 MP	T32TLR	24
Sodankylä peatland	67.37° N, 26.65° E	Stardot Netcam SC	5.0 MP	T35WMQ	22
Sodankylä canopy	67.36° N, 26.64° E	Stardot Netcam SC	5.0 MP		22
Lompolojankka peatland	69.80° N, 24.21° E	Stardot Netcam SC	5.0 MP	T34WFA	23
Kenttärova canopy	67.99° N, 24.24° E	Stardot Netcam SC	5.0 MP		23

2.3.2. In-Situ Snow Measurements

In Turkey, binary snow maps derived from S-2 imagery have been validated against ground-based measurements for the winter season 2017/18. Snow data from automatic weather stations (i.e., AWOS: Automated Weather Observing System, and SPA: Snow Pack Analyser) operated by Turkish state meteorological service (TSMS) have been used. Daily snow depth (SD) values have been obtained by processing and filtering the raw data supplied by these stations (e.g., removal of possible false snow detection due to grass). This analysis relies on SD measurements provided by 75 ground stations and 205 S-2 images available between November 2017 and April 2018 over Turkey. The validation has been performed over 25 S-2 tiles and 286 in-situ SD observations have been analyzed. Relative positions of S-2 tiles and the ground stations are shown in Figure 8.

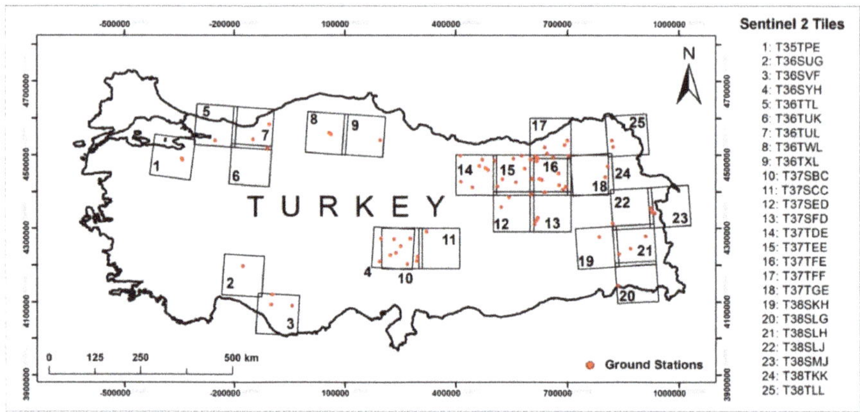

Figure 8. Locations of ground-based monitoring stations in S-2 tiles.

3. Methods

After introducing the retrieval algorithms implemented for the generation of the satellite snow products investigated in this study, the processing of in-situ data and imagery used to validate S-2 observations is described. The methodologies for the S-2-based assessment of H-SAF snow products and the assessment of S-2 imagery against in-situ data are presented and discussed, as well as the evaluation metrics.

3.1. Satellite Retrieval Algorithms

Snow detection can be retrieved from optical imagery through different algorithms, which generally rely on thresholding methods based on channel differences and ratios to exploit the different spectral properties of snow-covered areas with respect to snow-free surfaces and clouds. One of the most commonly-used indices is the normalized difference snow index (NDSI), which is defined as the difference of reflectance observed in a visible band and a shortwave infrared one, divided by the sum if the two reflectance values [72]. Indeed, since snow reflectance is high in the visible wavelengths and low in the shortwave infrared ones, this method enables distinguishing snow from clouds and other non-snow-covered conditions [72,73]. However, it is noteworthy that the suitability of each retrieval algorithm necessarily depends on the main features of the satellite data to be processed [55,74,75].

3.1.1. Sen2Cor Algorithm

S-2 L1C data are downloaded from the Copernicus open access hub. The L1C image product consists of a series of 100×100 km^2-tiles, each of which is made of thirteen compressed JPEG-2000 images, one for every single band. The MSI TOA reflectance images are processed through the Sen2Cor version 2.5.5, namely the last version of Sentinel-2 L2A prototype processor provided by ESA. Sen2Cor consists of ten main modules and it can perform the tasks of atmospheric, terrain and cirrus correction of L1C input data to generate optimally corrected BOA reflectance images. In this study the L2A_SceneClass (SC) module is used to perform the classification of the input images and to generate Scene Classification (SCL) maps at a spatial resolution of 20 m. The SC algorithm allows the detection of clouds, snow and cloud shadows, and the generation of a classification map consisting of four different classes for clouds (including cirrus), together with six different classifications for shadows, cloud shadows, vegetation, soils/deserts, water and snow (Table 6). The SC module consists of the cloud/snow algorithm, the cirrus detection algorithm, and the cloud showdown detection algorithm to generate the classification map [76]. Each algorithm processes the TOA reflectance input data through a sequence of thresholding filters, which are applied to S-2 spectral bands, band ratios, and indexes. In the cloud/snow detection algorithm, each test provides a cloud probability, which is recursively updated at each step. After thresholding the brightness in the red region of the solar spectrum (band 4), all potentially cloudy pixels are filtered by thresholding the NDSI [72], which is evaluated from spectral bands 3 and 11. Snow confidence map is generated by detecting snow pixels, according to four successive filters using spectral bands 2, 3, 8, 11. Ancillary information on yearly snow climatology is used to define the monthly snow probability of each pixel and to discard possible false snow detections. All potentially snow pixels are then filtered by sequentially thresholding the reflectance in band 8 (NIR) and band 2 (blue), and the ratio between band 2 and band 4 to identify main water bodies. Lastly, possible false cloud detection at the boundaries of snowy regions is removed by performing a brightness test on band 12. Once cloud and snow confidence masks are generated, an optional spatial filter can be applied to reduce possible false cloud detection. The cirrus detection algorithm mainly relies on the reflectance thresholding of band 10, because of the high water vapor absorption in this region [77], and an additional cross check is performed against the probabilistic cloud mask.

Table 6. Classes of Sen2Cor SCL map.

Label	Classification
0	No data
1	Saturated/defective
2	Dark area
3	Cloud shadows
4	Vegetation
5	Not vegetated
6	Water
7	Unclassified
8	Cloud (medium probability)
9	Cloud (high probability)
10	Thin cirrus
11	Snow

3.1.2. H-SAF H10 Algorithms

The distribution of snow cover and the non-uniformity of snow properties are significantly different over mountainous and flat/forested areas, since they strongly depend on the local topography and vegetation cover. Therefore, two distinct stand-alone algorithms are implemented within the product generation and applied according to a mountain mask defined in [56]. The algorithm for flat/forested areas has been developed by Finnish Meteorological Institute (FMI) [70,78]. The algorithm utilizes TOA radiances of six SEVIRI channels (0.635, 0.81, 1.64, 3.90, 10.80, 12.00 μm), the brightness temperatures of three channels (3.90, 10.80, and 12.00 μm), sun and satellite zenith and azimuth angles, the International Geosphere-Biosphere Programme (IGBP) land-cover type by the U.S. Geological Survey (USGS), and the land surface temperature (LST) classification produced by the EUMETSAT's Satellite Application Facility on Land Surface Analysis (LSA SAF) [77]. While information from channels around 0.635, 0.81, and 1.64 μm are used to classify different surface types [72,79,80], the algorithm exploits the radiance ratio of SEVIRI channels 2 (0.81 μm) and 3 (1.64 μm), and the brightness temperature difference of channels 10 (12.00 μm) and 4 (3.90 μm) to properly detect clouds [81,82]. The Middle East Technical University (METU) developed the algorithm for mountainous areas [56], which exploits 4 SEVIRI spectral channels (0.635 μm, 1.64 μm, 3.90 μm, 10.80 μm). Cloud discrimination is preliminary performed to identify cloud-free pixels by jointly using Cloud Mask (CMa) and Cloud Type (CT) products of the EUMETSAT's Nowcasting Satellite Application Facility (NWC SAF) [83]. Firstly, pixels having reflectance values higher than 0.35 are collected, because of the high visible reflectance of snow. Secondly, since snow cover has a low reflectance in the middle infrared and a high reflectance in the visible, pixels having snow index (SI) value lower than 0.6 are collected, which is evaluated by dividing channel 3 (1.64 μm) to channel 1 (0.635 μm) [84,85]. Lastly, pixels having temperature lower than 288 K on channel 9 (10.80 μm) are accepted, considering that the temperature of snow cannot exceed the freezing point [86]. It is noteworthy that sun zenith angle (SZA) thresholds are applied, SZA > 80 in the FMI algorithm and SZA > 85 in the METU algorithm are used for discarding the low-illuminated areas. No atmospheric correction is included in both algorithms. The final snow recognition product results from the merging of the products for flat/forested and mountainous areas over the full H-SAF spatial domain.

3.1.3. H-SAF H12 Algorithms

Consistently with H-SAF H10 product, two different retrieval algorithms are separately applied for flat/forested and mountainous areas. Since the observing cycle of satellites over Europe is about 3 h, the scenes are multi-temporally analyzed to search for time instants of cloud-free conditions in 24 h. The retrieval algorithm of FSC in forested/flat areas has been developed at FMI. The method is based on a semi-empirical reflectance model [87], which evaluates the reflectance as a function of the snow-covered area by using visible and near-infrared data (visible band 1) [88]. Since forest transmissivity is of critical importance to estimate the snow-covered area in all conditions of forest coverage, the algorithm relies on the transmissivity map generated from reflectance data acquired at full dry snow cover conditions to guarantee a proper contrast between forest canopy and ground. However, a priori information on forests is not needed, because the effective average forest transmissivity is estimated from Earth observation reflectance data. Conversely, the retrieval algorithm for mountainous areas involves the thresholding of NDSI at 0.4, since it allows the derivation of the resulting average fraction of snow-covered area by retrieving snow and snow-free ground from satellite data according to the reflectance values [55]. Because in mountainous regions the sun zenith and azimuth angles, as well as direction of observation relative to these are the most limiting factors [89], possible terrain effects are properly removed from the measured radiance through a statistical-empirical correction method [90]. Once the visible channel is corrected due to topographic effect, the average reflectance values are determined from pixels of pure snow-covered area and pure bare ground. In defining the pure snow-covered area and snow-free bare ground, the NDSI threshold greater than 0.4 and lower than 0 are used, respectively. The original model has previously been developed by [91]. According to this approach, the pixel reflectance is modelled as a linear mixture of snow, individual tree species and snow-free bare ground (e.g., rock, soil, low vegetation). For mountainous areas the equation considers snow and bare ground, because of the general lack of trees at high altitudes:

$$\overline{R}_G = A_{SW} R_{SW} + A_{BG} R_{BG} \tag{1}$$

where $\overline{R}_G$ is the modelled pixel reflectance for a given wavelength, A represents area fractions of a pixel (with $A_{SW} + A_{BG} = 1$), R is the reflectance, subscripts SW and BG refer to snow and bare ground, respectively.

It is noteworthy that both atmospheric and topographic corrections are implemented in H12 algorithms. The products for flat/forested and mountainous areas are merged over the full H-SAF area, according to the mountain mask defined in [56]. The merging algorithm is properly designed to minimize the projection errors [92].

3.2. Validation of Sentinel-2 Imagery with In-Situ Data

3.2.1. Validation of Sentinel-2 Imagery by In-Situ Webcams

The validation is based on the comparison of single daily FSC values derived from camera observations to the corresponding single values obtained from S-2-derived FSC maps over the observed area. FSC values have been estimated by experts through the visual inspection of camera images. Visual inspections have been limited over area of interests (AOIs) selected according to both camera properties and the local topographic features, so that the snow cover is clearly visible, and the relative surface area can be estimated as accurately as possible. FSC values have been estimated in 10%-intervals (i.e., 0%, 10%, ... , 90%, 100%). The visual inspection of each image has been performed by 4 expert observers. With the aim of assessing the subjective error, the resulting RMSE of FSC estimates has been also calculated, as shown in Table 7. The study of Arslan et al. (2017) [33] has estimated that the subjective error is within 10%, in terms of FSC. For the comparison, average values of the visual estimates have been used to minimize the subjective error.

Table 7. AOI Sizes, corresponding number of S-2 pixels and subjective error of webcam data observers.

Site Name	AOI Size [m²]	Number of S-2 Pixels	RMSE (All Days)	RMSE (Only Patchy Snow Cover)
Torgnon	1,056,171	2722	13.6%	13.6%
Sodankylä peatland	3976	9	0%	0%
Sodankylä canopy	4760	11	6.3%	13.2%
Lompolojankka peatland	12,310	33	5.7%	15.8%
Kenttärova canopy	254,373	633	0%	0%

To properly validate the mapping of S-2-derived FSC, a mask for each selected webcam has been created by drawing polygons over the approximate AOIs using Google Earth. This has been done by visually comparing the landmarks in webcam images and Google Earth optical satellite data overlay. AOIs have been modified according to the landmarks so that the polygons were as accurate as possible. The polygons have been then converted into GeoTIFF files to be used in the FSC evaluation over each AOI from S-2-based snow cover maps (S-2-derived FSC). Area sizes of those polygons and the corresponding number of S-2 pixels are shown in Table 7. The comparison has been performed also over the three AOIs corresponding to relatively low numbers of pixels in S-2 grid, since in those sites most of the images were available either in full snow cover or snowless conditions, which make this analysis feasible. Along with FSC, cloud cover fraction is also calculated for each AOI.

As an example, AOI for the Kenttärova canopy camera and S-2 derived snow cover map over the AOI for 18 February 2018 are shown in Figure 9. In Figure 9c are reported the approximate camera FOV (white polygon) and the selected AOI (yellow polygon) in the Google Earth view. AOI from the camera FOV is reported in Figure 9a,b. The snow cover map derived from the S-2 image over the AOI polygon is shown in Figure 9d. In this example, while experts have estimated FSC as full snow cover (100%) from camera images (at 09:00 and 11:00), S-2-derived FSC (at 10:10) has been estimated as 82%. However, it is noteworthy that 38% of the analysed pixels over the AOI are cloudy or unclassified.

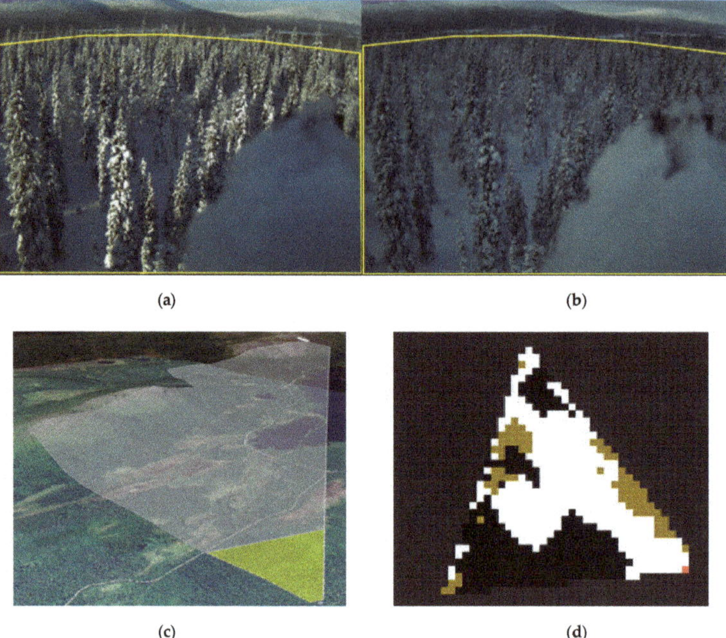

Figure 9. AOI for Kenttärova canopy camera and S-2 snow cover map over the AOI for 18 February 2018–(**a**) AOI marked on the camera image at 09:00 (yellow polygon); (**b**) AOI marked on the camera image at 11:00 (yellow polygon); (**c**) Approximate FOV of the camera (white polygon) and AOI (yellow polygon) in Google Earth; (**d**) Extracted snow cover map from S-2 image at 10:10 (white: snow; brown: no-snow; black: clouds and unclassified; red: camera location). AOI is approximately 0.25 km^2, corresponding to 633 S-2 pixels.

After obtaining the value pairs for the comparison, the ones having cloud cover fraction over 50% have been filtered out. The value pairs of FSC have been compared and the resulting RMSE values have been evaluated.

3.2.2. Validation of Sentinel-2 Imagery against Ground-Based Snow Measurements

The procedure to validate S-2 snow mapping against ground-based data relies on the thresholding of SD measurements to properly define snow and snowless conditions. According to the in-situ measures, the presence of snow is detected whenever a threshold of 5 cm is exceeded. This threshold has been set due to the expected uncertainty in measuring devices [93].

3.3. Procedures of Cross-Sensor Comparison between Satellitesnow Products

After processing S-2 data through Sen2Cor SC module, a quality check of the satellite data series has been performed through random visual inspections to prevent possible systematic inconsistencies. The comparison between observations sensed by different sensors on the same day is performed at the scale of S-2 tile. Indeed, the consistency assessment of H-SAF products against S-2 data is carried out individually over each single tile. In order to properly perform the comparison analysis, all the satellite snow products have been preliminarily re-projected to the same common image projection, namely WGS84/UTM. Since the analysis is tile-based, maps over the geographic extension of each selected S-2 tile are derived from the original full images of both H-SAF products. The selection of the data subset limited over the domain of each tile is carried out by considering the local coordinates of tiles borders, in order to properly guarantee the intersection of the satellite products over each S-2 tile.

3.3.1. Comparison between Sentinel-Based Snow Masks and H-SAF H10

Figure 10 shows the comparison procedure. Firstly, binary snow masks (presence/absence of snow cover) are derived from both H10 and S-2 SCL maps. According to the original classification of SCL map (Table 6), vegetation, not vegetated, and water (Table 6, classes 4, 5, 6) pixels have been classified as no-snow pixels. Unclassified (Table 6, classes 0, 1, 2, 7) and cloud-contaminated (Table 6, classes 3, 8, 9, 10) pixels are flagged and neglected in the comparison of snow maps, with the aim of preventing possible cloud cover affecting the snow detection [57]. Consistently, H10-based snow masks are derived by considering snow and bare-soil pixels. Since the satellite products are differently gridded, the comparison is performed at the coarser spatial resolution of the H-SAF H10 [55]. For each H10 grid cell, the percentage of snow cover is determined according to the S-2 observations by counting the number of S-2 snow pixels versus the total number of S-2 pixels in the coarser cell [55]. This computation results into an S-2-based FSC map (S-2-derived FSC). In order to restore a binary snow mask, each resulting S-2-based coarse cell is then classified as snow if FSC is higher than 50%, otherwise it is classified as soil [57,94]. S-2-based coarse cell where more than the 50% of fine S-2 pixels are classified as cloud or unclassified are neglected in order not to compromise the analysis results. A preliminary analysis has been performed by testing threshold values equal to 25%, 50%, 75% to properly assess the impact of the thresholding of cloud cover at pixel scale. The results have revealed a poor sensitivity of the comparison procedure to the threshold value.

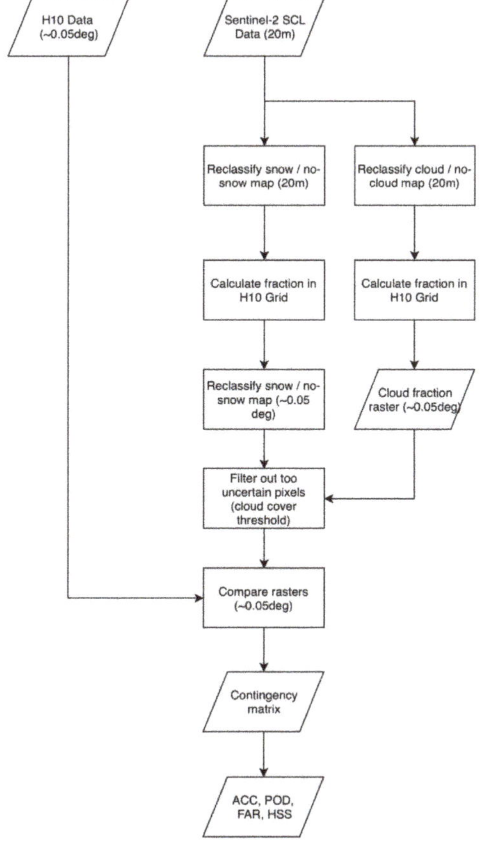

Figure 10. Flowchart of comparison procedure of H10 product.

3.3.2. Comparison between Sentinel-Derived FSC Maps and H-SAF H12

Consistently with the assessment of H-SAF H10 product, the analysis of H12 data relies on the same procedure and assumptions. The only main difference is the lack of the thresholding of FSC derived from S-2 imagery over the coarser H-SAF grid (S-2-derived FSC), since it is directly comparable with H12 product. Indeed, this analysis compares FSC maps of H12 product, which are generated through retrieval algorithms (Section 3.1.3), with the S-2-based FSC maps, which are derived by counting the number of S-2 snow pixels versus the total number of S-2 pixels in the coarser H12 cell. The comparison scheme is reported in Figure 11.

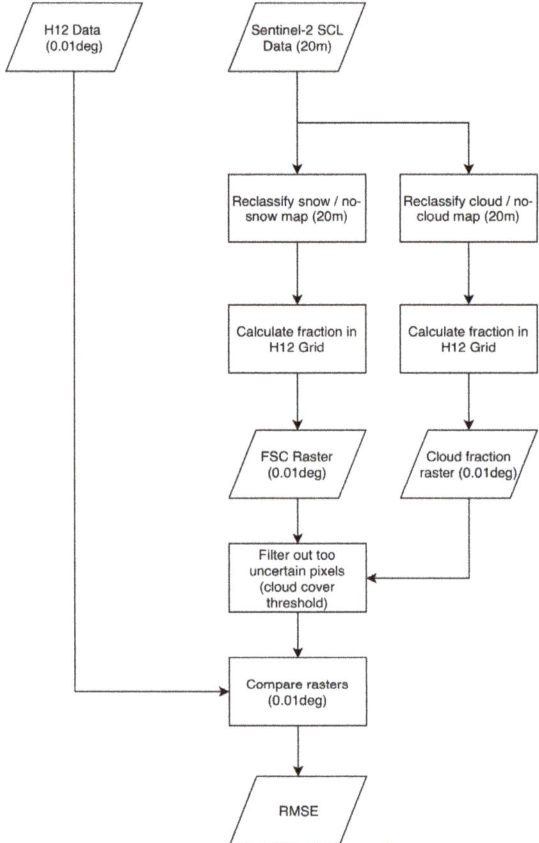

Figure 11. Flowchart of comparison procedure for H12 product.

It is noteworthy that when mapping the snow cover through remotely-sensed optical imagery, forests constitute a challenge, since the canopy (1) partially obscures the signal along the path from ground to sensor and, (2) alters the observed reflectance [87]. Therefore, the accuracy of snow mapping generally decreases in forested areas with respect to non-forested regions [72,95,96]. Even though several methodologies have been proposed for the detection of snow under forest canopy [87,91,96–100], this issue still remains a critical research topic. Subpixel classification methods [87,101,102] are used to generate FSC maps with the aim of overcoming the limitations related to mixed-pixels problem affecting coarse-resolution imagery, namely possible mixtures of land cover classes (i.e., snow, soil, rock, vegetation, water, etc.) and area fractions of different cover classes within a pixel.

When assessing the consistency between S-2 and H12 data, the impact of vegetation on snow detection need to be investigated. Indeed, since the H12 retrieval algorithm in flat/forested area involves transmissivity maps, this product results from snow detection at ground level. Conversely, S-2 snow mask is derived from snow detection on canopy and thus it can be hindered by the presence of vegetation, mainly where forest cover is present. It is noteworthy that this difference in retrieval algorithms is supposed not to affect the analysis shortly after snowfall events, when forested areas are likely to be classified as snow pixels because of canopy interception [103]. On the other hand, during periods when no snowfall event occurs, the comparison between the two snow products can be more challenging. Indeed, especially in dense forests, the lack of intercepted snow can lead to a misleading S-2 classification as snow-free surface despite the presence of snow cover under canopy. Therefore, with the aim of addressing this critical issue, a further analysis has been performed according to the information on different vegetation types supplied by ESA GlobCover 2009 land cover map. The impact of the vegetation cover has been investigated throughout the whole analysis period by considering a sample of one tile in each country. The comparison between S-2-derived FSC and H12 product has been performed after preliminarily filtering out of the vegetated pixels in S-2 data by using the information supplied by GlobCover data for flat areas and without filtering in mountainous regions. For the filtering, S-2 derived FSC maps have been collocated with the GlobCover map. The pixels corresponding to flat regions in the mountain mask and belonging to the vegetation class V1 in GlobCover (described in Section 2.2) have been discarded from S-2 derived FSC maps. After that, the comparison has been performed through the same algorithm previously described. This procedure has also been applied for the vegetation class V2 to properly assess the impact of different vegetation types on snow detection. Results are presented separately for both classes.

3.4. Evaluation Metrics

For the consistency assessment of the mapping of snow extent, a contingency table is evaluated (Table 8).

Table 8. Contingency table reporting number of HITS (a), number of FALSE ALARMS (b), number of MISSES (c), number of CORRECT NEGATIVES (d).

		Reference Dataset	
		Snow	**No Snow**
Analyzed dataset	**Snow**	a	b
	No Snow	c	d

From these classification results, different scores for dichotomous statistics are evaluated:

- Probability of detection:

$$POD = a/(a + c) \tag{2}$$

- False alarm ratio:

$$FAR = b/(a + b) \tag{3}$$

- Probability of false detection:

$$POFD = b/(b + d) \tag{4}$$

- Accuracy:

$$ACC = (a + d)/(a + b + c + d) \tag{5}$$

- Critical success index:

$$CSI = a/(a + b + c) \tag{6}$$

- Heidke skill score:

$$\text{HSS} = 2(ad - bc)/[(a + c)(c + d) + (a + b)(b + d)] \tag{7}$$

FSC is assessed through the evaluation of Root Mean Square Error (RMSE) with respect to the reference dataset $\left(FSC_{ref} \right)$.

$$\text{RMSE} = \sqrt{\frac{\sum_{i=1}^{n} \left(FSC_{ref,i} - FSC_i \right)^2}{n}} \tag{8}$$

4. Results and Discussion

4.1. Validation of Sentinel-2 Imagery

When validating S-2 imagery against in-situ observations, it is noteworthy that the evaluation metrics have been evaluated by considering ground-based datasets as the reference ones.

4.1.1. In-Situ Digital Imagery

The comparison relies on a total of 50 pairs of FSC values, resulting from the analysis of matching webcam and S-2 images, both in Finland and in Italy. The evaluation has revealed a total RMSE value of 12.22%. In overall, S-2 snow mapping reveals a general FSC overestimation. When neglecting full-snow and bare-soil classification (i.e., FSC equal to 0% and 100%), the total RMSE value increases up to 19.82% for 19 value pairs. It is noteworthy that no outlier affects the distribution and the analyzed data boast a high correlation. Figure 12 shows the data scatterplot resulting from the comparison.

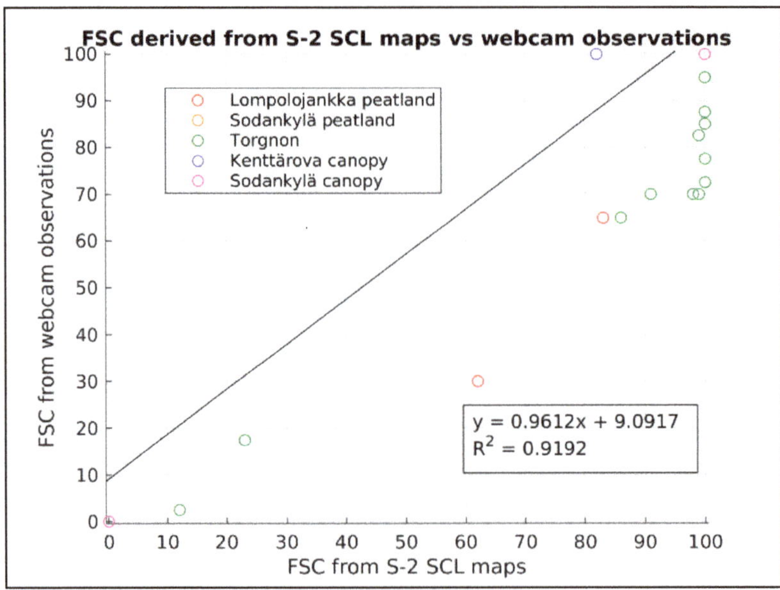

Figure 12. Distribution of compared value pairs.

The undesired cases resulting in high errors have been investigated in more detail. In Torgnon, where the AOI is selected over the mountainous area, the three scenes having the highest error are those affected by the largest cloud cover fraction, greater than 29%. Indeed, in the presence of patchy snow cover, partial cloud cover over the area is likely to unavoidably affect the FSC derived from S-2 data. Under conditions of patchy snow cover, S-2 data are affected by overestimation in Lompolojankka.

Two occurrences are during the melting period, where the ground is mostly cover by meltwater. One further occurrence is in early winter, where snow cover is still not full and sparse vegetation is likely to hinder the visual inspection.

4.1.2. Ground-Based Snow Measurements

The validation of S-2 snow mapping against ground-based SD measurements in Turkey has revealed a significant consistency of satellite imagery, as evidenced by the highest number of hits and lower values of false alarms and misses (Table 9).

Table 9. Contingency table of ground-based validation of S-2 binary snow maps in Turkey for winter season 2017/18.

		Ground-Based Measures	
		SD $\geq$ 5 cm	SD < 5 cm
S-2 Binary Snow Masks	**Snow**	201	17
	No Snow	43	25

As shown in Table 10 reporting the resulting evaluation metrics, the remotely-sensed high-resolution observations properly succeed in detecting the presence of snow cover.

Table 10. Evaluation metrics of ground-based validation of S-2 binary snow maps in Turkey for winter season 2017/18.

Metrics	Value
POD	0.82
FAR	0.08
POFD	0.40
ACC	0.79
CSI	0.77
HSS	0.33

4.2. Cross-Sensor Comparison of Snow Extent Products

For each S-2 tile a pixel-to-pixel analysis has been performed to evaluate the consistency between the S-2-based maps of snow extent and H-SAF H10 product.

Figures 13–15 show the comparison results in terms of POD, FAR and ACC for each analyzed tile in Italy, Finland and Turkey, respectively. When assessing the evaluation metrics, it is noteworthy that H10 product generally reveals higher performances over flat areas (i.e., Finland), rather than over mountainous regions (i.e., Italian Alps and Turkey). This issue is mainly due to the impact of the local complex topography affecting the sensors capability to detect snow. Indeed, mapping the high spatial variability of snow cover distribution over mountain sides is a challenging task at the coarser satellite resolution. Conversely, when considering the vegetation cover of each pixel (Table 3), even the presence of the vegetation species supposed to hinder the snow detection (e.g., evergreen needle-leaved forest) results in a lesser impact than topographic factors. However, both in Italy and Turkey, H10 product reveals a slightly weaker reliability over the most vegetated tiles, namely T32TQS and T33TUM, and T36TWL, respectively. This result suggests that the vegetation has a greater impact where the local topography is complex, due to overlapping effects. Nevertheless, H10 product generally ensures an accuracy greater than 0.8, except for tiles T32TNS and T33TPS over Italian Alps.

With the aim of investigating whether product performances are affected by the seasonality of snow cover, the evaluation metrics have been assessed under different snow cover conditions. Indeed, three different periods have been individually assessed, namely early winter (i.e., October and November), winter (i.e., December-March), melting period (i.e., April and May) (Figure 16).

Consistently with the tile-scale analysis, the agreement between S-2 and H10 data is higher over flat areas (i.e., in Finland). The analyses show that in early winter the lower accuracy and higher FAR values of H10 product over flat areas are mainly due to frequent cloudiness, which is likely to affect the snow detection. However, it is noteworthy that the presence of canopy is likely to have a higher impact during early winter and melting period. Indeed, in those months, the presence of patchy snow cover and the lower frequency of snowfall events intercepted by canopy make the snow mapping more challenging, especially where dense forests are present. Furthermore, it is important to consider that the 50%-thresholding of FSC derived from S-2 data (Section 3.2) is likely to affect the analysis mainly during the transition periods, when patchy snow cover is present.

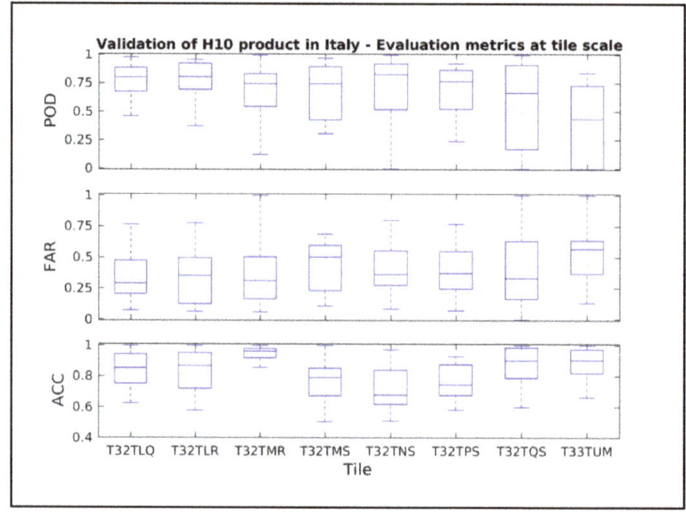

Figure 13. Evaluation metrics at tile scale in Italy.

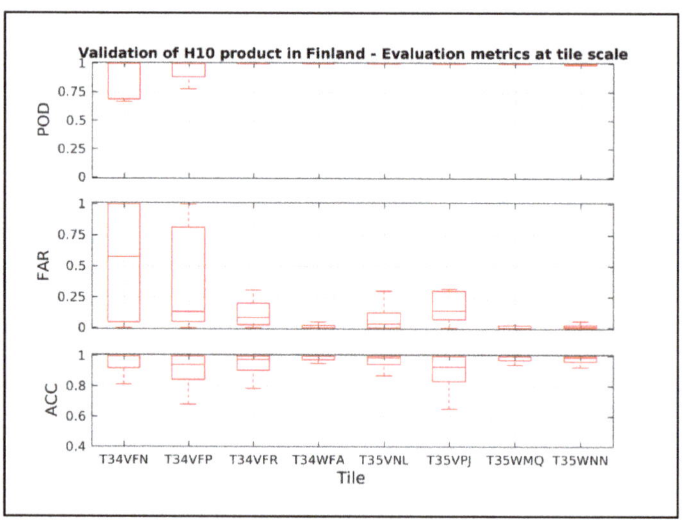

Figure 14. Evaluation metrics at tile scale in Finland.

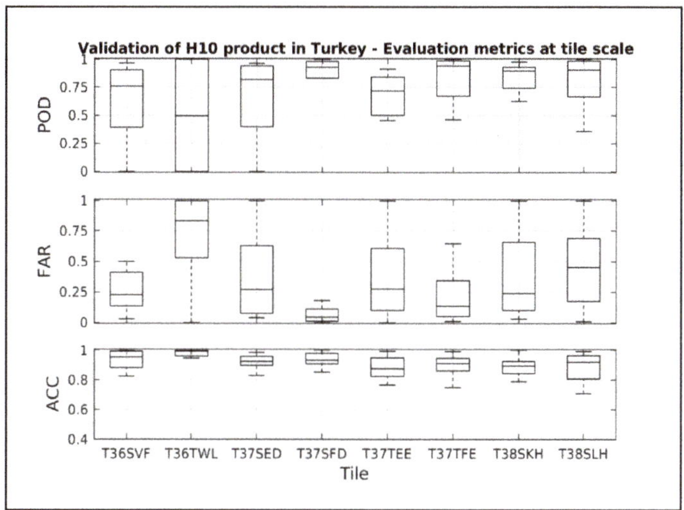

Figure 15. Evaluation metrics at tile scale in Turkey.

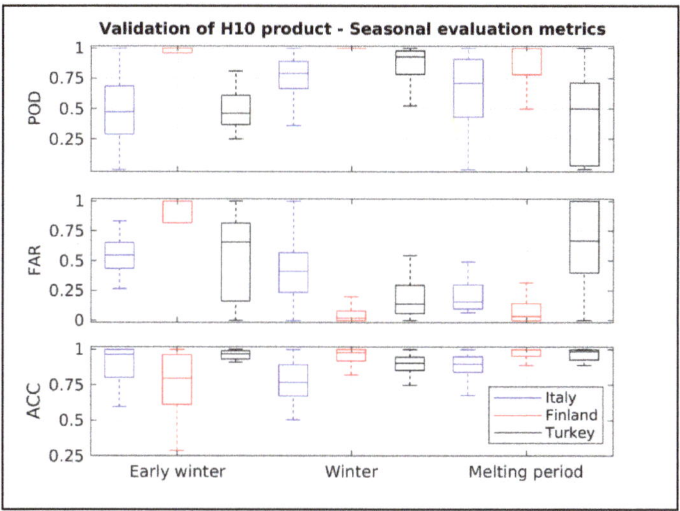

Figure 16. Seasonal evaluation metrics in Italy, Finland, and Turkey.

The higher consistency between H10 product and S-2 snow masks over flat areas is confirmed when assessing the evaluation metrics throughout the whole analysis period (Table 11). While H10 product reveals satisfying evaluation metrics of the same ranges in Finland and in Turkey, the Alpine complex topography strongly affects the snow mapping over this region, as proved by the poorer scores.

Table 11. Comparison between S-2-based snow masks and H10 product—Median values of evaluation metrics throughout the whole analysis period. H10 product requirements for both flat/forested (i.e., Finland) and mountainous areas (i.e., Italian Alps, Turkey) (POD_{thr} and FAR_{thr}) are reported [104].

Area	POD_{thr}	FAR_{thr}	POD	FAR	POFD	ACC	CSI	HSS
Finland	0.80	0.20	0.98	0.10	0.07	0.95	0.89	0.90
Italian Alps			0.78	0.35	0.16	0.83	0.55	0.59
Turkey	0.60	0.30	0.91	0.13	0.08	0.92	0.80	0.83

4.3. Cross-Sensor Comparison of Effective Snow Cover Products

The agreement between the mapping of the FSC derived from S-2 imagery and H12 product has been assessed according to the same pixel-to-pixel approach.

Like in the assessment of H10 data, the complex topography in mountainous areas affects the consistency between the two analyzed datasets, especially over the Italian Alps, where RMSE values are higher than the other case studies (Figure 17). However, RMSE scores are generally lower than 0.4, except for the same tiles in Italy, in compliance with the product requirements [105].

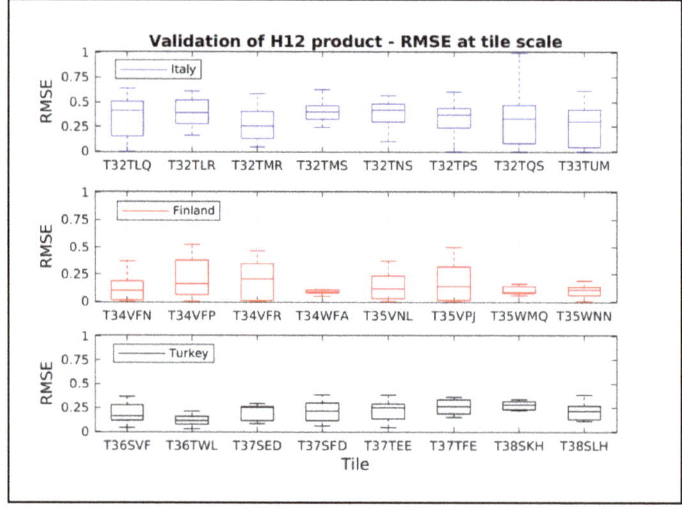

Figure 17. RMSE at tile scale in Italy, Finland, and Turkey.

Figure 18 shows that during the winter period RMSE values are generally higher than in other seasons. This issue is mainly due to overestimated classifications as full snow cover over the coarser spatial resolution of H12 product with respect to that derived from 20-m S-2 imagery, especially in mountainous regions.

The RMSE assessment over the whole analysis period confirms higher performances of H12 product over flat areas (i.e., Finland) than in mountainous regions (i.e., Italian Alps and Turkey) (Table 12). However, the product target requirements [105] are generally satisfied over both mountainous (RMSE ~ 30%) and flat (RMSE ~ 20%) areas.

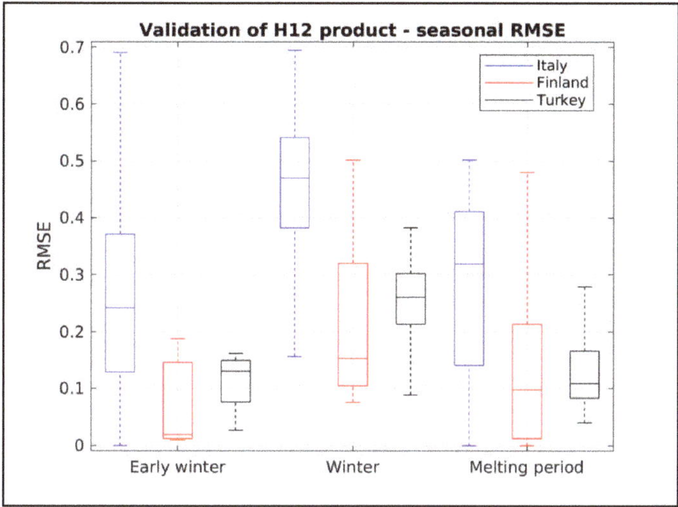

Figure 18. Seasonal RMSE in Italy, Finland, and Turkey.

Table 12. Overall RMSE values for the comparison between S-2-derived FSC maps and H12 product. H12 product requirement for both flat (i.e., Finland) and mountainous areas (i.e., Italian Alps, Turkey) ($RMSE_{thr}$) is reported [105].

Region	$RMSE_{thr}$	RMSE
Finland	0.40	0.15
Italian Alps	0.50	0.33
Turkey		0.21

Impact of Vegetation on Snow Detection

In order to properly assess the impact of the vegetation cover within the assessment of H12 product, the results obtained by filtering out the main classes V1 and V2 are individually evaluated against the reference ones relying on all S-2 pixels. As expected, the vegetation cover has a negligible impact on the comparison results in the Alpine region, since H12 and S-2 snow retrieval algorithms are consistent over the mountain mask (Section 3.2.2). Conversely, in Finland and Turkey the comparison procedure reveals a slight sensitivity to the different vegetation classes. While the snow detection is more affected by V1-vegetation class (i.e., needle-leaved evergreen forest) in Finland, V2-vegetation class (i.e., broadleaved deciduous forest) has a higher impact in Turkey. Figure 19 shows that by filtering out V1-vegetation pixels, the RMSE value in Finland increases with respect to the all-pixels analysis. The reduction in consistency between H12 and S-2 snow mapping suggests a good agreement of the two datasets over dense forests mainly due to the long-lasting snow interception over canopy during the winter season. Consistently, in Turkey the resulting RMSE slightly increases when filtering V2-vegetation pixels.

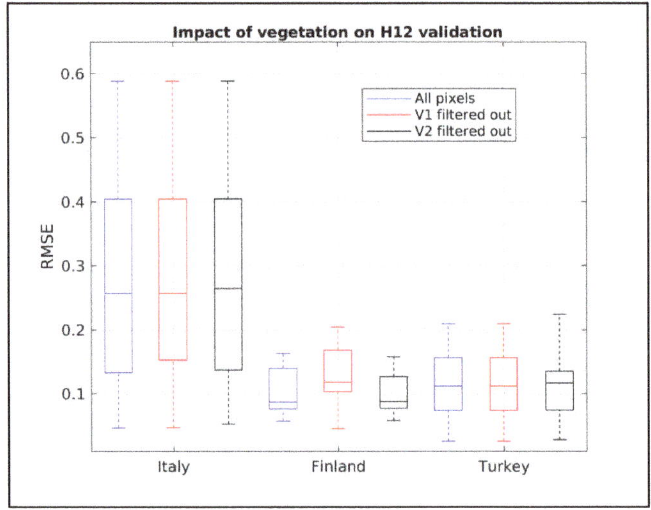

Figure 19. Impact of vegetation types on the comparison procedure of H12 product.

5. Conclusions

This study investigated the potential of using S-2 data to assess moderate-resolution EUMETSAT H-SAF product of snow extent (H10) and FSC (H12) in Finland, the Italian Alps and Turkey. Snow masks derived from S-2 imagery have revealed a significant consistency with both ground-based snow measurements (POD = 0.82, FAR = 0.08,) and in-situ webcam photography, revealing a RMSE of about 12%, in terms of FSC. Hence the reliability of assuming this high-resolution dataset as a reference for intercomparison purposes. The results obtained in this study reveal that S-2 data can be properly used to continuously assess these medium resolution satellite snow products, which have been commonly validated against in-situ data so far [53,56]. However, it is noteworthy to consider that under specific conditions the snow mapping derived from S-2 data can be affected by critical flaws. Indeed, the analysis of camera images in the Italian Alps has shown that dense cloud cover can undermine the reliability of S-2 snow masks, mainly when patchy snow cover is present. Furthermore, during melting period the widespread presence of meltwater over flat areas may lead to an overestimation of snow cover.

The results of the cross-sensor comparison prove that the analyzed H-SAF snow products are highly consistent with S-2 imagery in detecting snow, also in terms of FSC, generally in compliance with the products requirements [104,105]. Nevertheless, the analyzed satellite datasets generally reveal a higher agreement over flat/forested areas (POD_{H10} and $RMSE_{H12}$ equal to 0.98 and 0.15, respectively) than in mountainous regions (over Italian Alps, POD_{H10} and $RMSE_{H12}$ equal to 0.78 and 0.33, respectively). Indeed, the local complex topography is likely to significantly hinder snow detection over mountain sides at coarser satellite spatial resolution. Conversely, the vegetation cover has turned out to have a less relevant impact on the consistency among remotely-sensed observations, even in the presence of dense evergreen forest. In Finland the long-lasting snow interception on vegetation canopy is expected to contribute to strengthening the agreement between S-2 snow maps and H12 images during the winter season. However, further key issues need to be addressed in the future. Primarily, a comparative study on different retrieval algorithms would allow an assessment of the reliability of the snow mapping derived from S-2 imagery. Secondly, the impact of cloudiness on the consistency among remotely-sensed observations should be investigated in more details through the analysis of scenes with different cloud cover percentages.

Nevertheless, these promising results currently encourage the effective use of the analyzed H-SAF snow products for hydrological and climatological studies, since they provide reliable snow-related information at large scale. Furthermore, thanks to their free availability at a daily scale, both H10 and H12 products are recommended as particularly suited for operational applications.

Author Contributions: All co-authors contributed to the design of this study. Validation algorithms were developed by C.M.T., G.P., A.N.A., Z.A., S.K., B.S., S.G., and M.T. Collection and validation of satellite data were performed by C.M.T. and B.S. in Finland, G.P. and C.M.T. in Italy, Z.A. and S.K. in Turkey. G.P. wrote the paper with contributions of C.M.T., S.K. The manuscript was reviewed by A.N.A., Z.A., C.M.T., S.K., S.G., B.S.

Funding: This research received no external funding.

Acknowledgments: The research has been conducted in the framework of the EUMETSAT H-SAF Project, thanks to the collaboration among several partner institutes of the validation cluster of snow products. We wish to thank all the institutions for supporting and promoting these research activities. We are primarily grateful to the European Cooperation in Science and Technology (COST), which has funded a Short Term Scientific Mission (STSM) of key importance for the success of this collaboration through the ES1404 Harmosnow Action. The work has been also supported by the Italian National Department of Civil Protection. The authors would like to acknowledge the Environmental Protection Agency of Aosta Valley (Italy) for providing Italian webcam datasets used in this study. We would like to address special thanks to Edoardo Cremonese for the fruitful discussions on this matter.

Conflicts of Interest: The authors declare no conflict of interest.

References

1. Bates, B.; Kundzewicz, Z.; Wu, S. *Climate Change and Water*; Intergovernmental Panel on Climate Change Secretariat: Geneva, Switzerland, 2008.
2. Appel, I. Uncertainty in satellite remote sensing of snow fraction for water resources management. *Front. Earth Sci.* **2018**, *12*, 711–727. [CrossRef]
3. Thirel, G.; Salamon, P.; Burek, P.; Kalas, M. Assimilation of MODIS snow cover area data in a distributed hydrological model using the particle filter. *Remote Sens.* **2013**, *5*, 5825–5850. [CrossRef]
4. Kumar, S.V.; Peters-Lidard, C.D.; Arsenault, K.R.; Getirana, A.; Mocko, D.; Liu, Y. Quantifying the added value of snow cover area observations in passive microwave snow depth data assimilation. *J. Hydrometeorol.* **2015**, *16*, 1736–1741. [CrossRef]
5. Aalstad, K.; Westermann, S.; Schuler, T.V.; Boike, J.; Bertino, L. Ensemble-based assimilation of fractional snow-covered area satellite retrievals to estimate the snow distribution at Arctic sites. *Cryosphere* **2018**, *12*, 247–270. [CrossRef]
6. Pirazzini, R.; Leppänen, L.; Picard, G.; Lopez-Moreno, J.I.; Marty, C.; Macelloni, G.; Kontu, A.; von Lerber, A.; Tanis, C.M.; Schneebeli, M.; et al. European In-Situ Snow Measurements: Practices and Purposes. *Sensors* **2018**, *18*, 2016. [CrossRef] [PubMed]
7. López-Moreno, J.I.; Nogués-Bravo, D. A generalized additive model for the spatial distribution of snowpack in the Spanish Pyrenees. *Hydrol. Process.* **2005**, *19*, 3167–3176. [CrossRef]
8. Bormann, K.J.; Westra, S.; Evans, J.P.; McCabe, M.F. Spatial and temporal variability in seasonal snow density. *J. Hydrol.* **2013**, *484*, 63–73. [CrossRef]
9. Luce, C.H.; Lopez-Burgos, V.; Holden, Z. Sensitivity of snowpack storage to precipitation and temperature using spatial and temporal analog models. *Water Resour. Res.* **2014**, *50*, 9447–9462. [CrossRef]
10. Rice, R.; Bales, R.C.; Painter, T.H.; Dozier, J. Snow water equivalent along elevation gradients in the Merced and Tuolumne River basins of the Sierra Nevada. *Water Resour. Res.* **2011**, *47*. [CrossRef]
11. Molotch, N.P.; Meromy, L. Physiographic and climatic controls on snow cover persistence in the Sierra Nevada Mountains. *Hydrol. Process.* **2014**, *28*, 4573–4586. [CrossRef]
12. Revuelto, J.; López-Moreno, J.I.; Azorin-Molina, C.; Vicente-Serrano, S.M. Topographic control of snowpack distribution in a small catchment in the central Spanish Pyrenees: Intra-and inter-annual persistence. *Cryosphere* **2014**, *8*, 1989–2006. [CrossRef]
13. Harpold, A.A.; Guo, Q.; Molotch, N.; Brooks, P.D.; Bales, R.; Fernandez-Diaz, J.C.; Musselman, K.N.; Swetnam, T.L.; Kirchner, P.; Meadows, M.W.; et al. LiDAR-derived snowpack data sets from mixed conifer forests across the Western United States. *Water Resour. Res.* **2014**, *50*, 2749–2755. [CrossRef]

14. Szczypta, C.; Gascoin, S.; Houet, T.; Hagolle, O.; Dejoux, J.F.; Vigneau, C.; Fanise, P. Impact of climate and land cover changes on snow cover in a small Pyrenean catchment. *J. Hydrol.* **2015**, *521*, 84–99. [CrossRef]

15. Fayad, A.; Gascoin, S.; Faour, G.; López-Moreno, J.I.; Drapeau, L.; Le Page, M.; Escadafal, R. Snow hydrology in Mediterranean mountain regions: A review. *J. Hydrol.* **2017**, *551*, 374–396. [CrossRef]

16. Gascoin, S.; Lhermitte, S.; Kinnard, C.; Bortels, K.; Liston, G.E. Wind effects on snow cover in Pascua-Lama, Dry Andes of Chile. *Adv. Water Resour.* **2013**, *55*, 25–39. [CrossRef]

17. Vionnet, V.; Martin, E.; Masson, V.; Guyomarc'h, G.; Bouvet, F.N.; Prokop, A.; Durand, Y.; Lac, C. Simulation of wind-induced snow transport and sublimation in alpine terrain using a fully coupled snowpack/atmosphere model. *Cryosphere* **2014**, *8*, 395–415. [CrossRef]

18. López-Moreno, J.I.; Fassnacht, S.R.; Heath, J.T.; Musselman, K.N.; Revuelto, J.; Latron, J.; Morán-Tejeda, E.; Jonas, T. Small scale spatial variability of snow density and depth over complex alpine terrain: Implications for estimating snow water equivalent. *Adv. Water Resour.* **2013**, *55*, 40–52. [CrossRef]

19. Raleigh, M.; Livneh, B.; Lapo, K.; Lundquist, J. How does availability of meteorological forcing data impact physically-based snowpack simulations? *J. Hydrometeorol.* **2016**, *17*, 99–120. [CrossRef]

20. Viviroli, D.; Archer, D.R.; Buytaert, W.; Fowler, H.J.; Greenwood, G.; Hamlet, A.F.; Huang, Y.; Koboltschnig, G.; Litaor, M.I.; López-Moreno, J.I. Climate change and mountain water resources: Overview and recommendations for research, management and policy. *Hydrol. Earth Syst. Sci.* **2011**, *15*, 471–504. [CrossRef]

21. Migliavacca, M.; Galvagno, M.; Cremonese, E.; Rossini, M.; Meroni, M.; Sonnentag, O.; Cogliati, S.; Manca, G.; Diotri, F.; Busetto, L.; et al. Using digital repeat photography and eddy covariance data to model grassland phenology and photosynthetic CO_2 uptake. *Agric. For. Meteorol.* **2011**, *151*, 1325–1337. [CrossRef]

22. Filippa, G.; Cremonese, E.; Migliavacca, M.; Galvagno, M.; Forkel, M.; Wingate, L.; Tomelleri, E.; di Cella, U.M.; Richardson, A.D. Phenopix: AR package for image-based vegetation phenology. *Agric. For. Meteorol.* **2016**, *220*, 141–150. [CrossRef]

23. Linkosalmi, M.; Aurela, M.; Tuovinen, J.-P.; Peltoniemi, M.; Tanis, C.M.; Arslan, A.N.; Kolari, P.; Aalto, T.; Rainne, J.; Laurila, T. Digital photography for assessing vegetation phenology in two contrasting northern ecosystems. *Geosci. Instrum. Methods Data Syst.* **2016**, *5*, 417–426. [CrossRef]

24. Peltoniemi, M.; Aurela, M.; Böttcher, K.; Kolari, P.; Loehr, J.; Karhu, J.; Linkosalmi, M.; Tanis, C.M.; Tuovinen, J.-P.; Arslan, A.N. Webcam network and image database for studies of phenological changes of vegetation and snow cover in Finland, image time series from 2014 to 2016. *Earth Syst. Sci. Data* **2018**, *10*, 173–184. [CrossRef]

25. Peltoniemi, M.; Aurela, M.; Böttcher, K.; Kolari, P.; Loehr, J.; Hokkanen, T.; Karhu, J.; Linkosalmi, M.; Tanis, C.M.; Metsämäki, S.; et al. Networked web-cameras monitor congruent seasonal development of birches with phenological field observations. *Agric. For. Meteorol.* **2018**, *249*, 335–347. [CrossRef]

26. Wingate, L.; Ogée, J.; Cremonese, E.; Filippa, G.; Mizunuma, T.; Migliavacca, M.; Moisy, C.; Wilkinson, M.; Moureaux, C.; Wohlfahrt, G.; et al. Interpreting canopy development and physiology using the EUROPhen camera network at flux sites. *Biogeosci. Discuss.* **2015**, *12*, 5995–6015. [CrossRef]

27. Richardson, A.D.; Hufkens, K.; Milliman, T.; Aubrecht, D.M.; Chen, M.; Gray, J.M.; Johnston, M.R.; Keenan, T.F.; Klosterman, S.T.; Kosmala, M.; et al. *PhenoCam Dataset v1. 0: Vegetation Phenology from Digital Camera Imagery, 2000–2015*; ORNL DAAC: Oak Ridge, TN, USA, 2017.

28. Farinotti, D.; Magnusson, J.; Huss, M.; Bauder, A. Snow accumulation distribution inferred from time-lapse photography and simple modelling. *Hydrol. Process.* **2010**, *24*, 2087–2097. [CrossRef]

29. Salvatori, R.; Plini, P.; Giusto, M.; Valt, M.; Salzano, R.; Montagnoli, M.; Cagnati, A.; Crepaz, G.; Sigismondi, D. Snow cover monitoring with images from digital camera systems. *Ital. J. Remote Sens.* **2011**, *43*, 137–145. [CrossRef]

30. Bernard, É.; Friedt, J.M.; Tolle, F.; Griselin, M.; Martin, G.; Laffly, D.; Marlin, C. Monitoring seasonal snow dynamics using ground based high resolution photography (Austre Lovenbreen, Svalbard, 79 N). *ISPRS J. Photogramm. Remote Sens.* **2013**, *75*, 92–100. [CrossRef]

31. Garvelmann, J.; Pohl, S.; Weiler, M. From observation to the quantification of snow processes with a time-lapse camera network. *Hydrol. Earth Syst. Sci.* **2013**, *17*, 1415–1429. [CrossRef]

32. Härer, S.; Bernhardt, M.; Corripio, J.G.; Schulz, K. PRACTISE-Photo Rectification and ClassificaTIon SoftwarE (V. 1.0). *Geosci. Model Dev.* **2013**, *9*, 307–321. [CrossRef]

33. Arslan, A.N.; Tanis, C.M.; Metsämäki, S.; Aurela, M.; Böttcher, K.; Linkosalmi, M.; Peltoniemi, M. Automated Webcam Monitoring of Fractional Snow Cover in Northern Boreal Conditions. *Geosciences* **2017**, *7*, 55. [CrossRef]

34. Tanis, C.; Peltoniemi, M.; Linkosalmi, M.; Aurela, M.; Böttcher, K.; Manninen, T.; Arslan, A. A System for Acquisition, Processing and Visualization of Image Time Series from Multiple Camera Networks. *Data* **2018**, *3*, 23. [CrossRef]

35. Gascoin, S.; Hagolle, O.; Huc, M.; Jarlan, L.; Dejoux, J.-F.; Szczypta, C.; Marti, R.; Sánchez, R. A snow cover climatology for the Pyrenees from MODIS snow products. *Hydrol. Earth Syst. Sci.* **2015**, *19*, 2337–2351. [CrossRef]

36. Nolin, A.W. Recent advances in remote sensing of seasonal snow. *J. Glaciol.* **2010**, *56*, 1141–1150. [CrossRef]

37. Frei, A.; Tedesco, M.; Lee, S.; Foster, J.; Hall, D.; Kelly, R.; Robinson, D. A review of global satellite-derived snow products. *Adv. Space Res.* **2012**, *50*, 1007–1029. [CrossRef]

38. Robinson, D.; Kukla, G. Maximum surface albedo of seasonally snow covered lands in the Northern Hemisphere. *J. Clim. Appl. Meteorol.* **1985**, *24*, 402–411. [CrossRef]

39. Nolin, A.W. Towards retrieval of forest cover density over snow from the Multi-angle Imaging SpectroRadiometer (MISR). *Hydrol. Process.* **2004**, *18*, 3623–3636. [CrossRef]

40. Derksen, C. The contribution of AMSR-E 18.7 and 10.7 GHz measurements to improved boreal forest snow water equivalent retrievals. *Remote Sens. Environ.* **2008**, *112*, 2701–2710. [CrossRef]

41. Dong, J.; Peters-Lidard, C. On the relationship between temperature and MODIS snow cover retrieval errors in the western US. *IEEE J. Sel. Top. Earth Obs. Remote Sens.* **2010**, *3*, 132–140. [CrossRef]

42. Maurer, E.P.; Rhoads, J.D.; Dubayah, R.O.; Lettenmaier, D.P. Evaluation of the snow-covered area data product from MODIS. *Hydrol. Process.* **2003**, *17*, 59–71. [CrossRef]

43. Tekeli, A.E.; Akyürek, Z.; Şorman, A.A.; Şensoy, A.; Şorman, A.Ü. Using MODIS snow cover maps in modeling snowmelt runoff process in the eastern part of Turkey. *Remote Sens. Environ.* **2005**, *97*, 216–230. [CrossRef]

44. Riggs, G.A.; Hall, D.K.; Salomonson, V.V. *MODIS Snow Products User Guide*; NASA Goddard Space Flight Center: Greenbelt, MD, USA, 2006.

45. Hall, D.K.; Riggs, G.A. Accuracy assessment of the MODIS snow products. *Hydrol. Process.* **2007**, *21*, 1534–1547. [CrossRef]

46. Akyurek, Z.; Sorman, A.U.; Sensoy, A.; Sorman, A.A. Calibration and Validation of satellite derived snow products with in situ data over the mountainous Eastern part of Turkey. In Proceedings of the International Congress on River Basin Management, Antalya, Turkey, 22–24 March 2007; pp. 711–726.

47. Parajka, J.; Blöschl, G. Spatio-temporal combination of MODIS images–potential for snow cover mapping. *Water Resour. Res.* **2008**, *44*. [CrossRef]

48. Wang, X.; Xie, H.; Liang, T.; Huang, X. Comparison and validation of MODIS standard and new combination of Terra and Aqua snow cover products in northern Xinjiang, China. *Hydrol. Process.* **2009**, *23*, 419–429. [CrossRef]

49. Huang, X.; Liang, T.; Zhang, X.; Guo, Z. Validation of MODIS snow cover products using Landsat and ground measurements during the 2001–2005 snow seasons over northern Xinjiang, China. *Int. J. Remote Sens.* **2011**, *32*, 133–152. [CrossRef]

50. Raleigh, M.S.; Rittger, K.; Moore, C.E.; Henn, B.; Lutz, J.A.; Lundquist, J.D. Ground-based testing of MODIS fractional snow cover in subalpine meadows and forests of the Sierra Nevada. *Remote Sens. Environ.* **2013**, *128*, 44–57. [CrossRef]

51. Arsenault, K.R.; Houser, P.R.; De Lannoy, G.J. Evaluation of the MODIS snow cover fraction product. *Hydrol. Process.* **2014**, *28*, 980–998. [CrossRef]

52. Byun, K.; Choi, M. Uncertainty of snow water equivalent retrieved from AMSR-E brightness temperature in northeast Asia. *Hydrol. Process.* **2014**, *28*, 3173–3184. [CrossRef]

53. Surer, S.; Parajka, J.; Akyurek, Z. Validation of the operational MSG-SEVIRI snow cover product over Austria. *Hydrol. Earth Syst. Sci.* **2014**, *18*, 763–774. [CrossRef]

54. Sönmez, I.; Tekeli, A.E.; Erdi, E. Snow cover trend analysis using interactive multisensor snow and ice mapping system data over Turkey. *Int. J. Climatol.* **2014**, *34*, 2349–2361. [CrossRef]

55. Salomonson, V.V.; Appel, I. Estimating fractional snow cover from MODIS using the normalized difference snow index. *Remote Sens. Environ.* **2004**, *89*, 351–360. [CrossRef]

56. Surer, S.; Akyurek, Z. Evaluating the utility of the EUMETSAT H-SAF snow recognition product over mountainous areas of eastern Turkey. *Hydrol. Sci. J.* **2012**, *57*, 1684–1694. [CrossRef]

57. Crawford, C.J. MODIS Terra Collection 6 fractional snow cover validation in mountainous terrain during spring snowmelt using Landsat TM and ETM+. *Hydrol. Process.* **2015**, *29*, 128–138. [CrossRef]

58. Metsämäki, S.; Ripper, E.; Mattila, O.P.; Fernandes, R.; Bippus, G.; Luojus, K.; Nagler, T.; Bojkov, B. Evaluation of Northern Hemisphere Snow Extent products within ESA SnowPEx-project. In Proceedings of the IEEE International Geoscience and Remote Sensing Symposium (IGARSS), Beijing, China, 10–15 July 2016; pp. 5280–5283.

59. Appel, I. Validation and potential improvements of the NPP fractional snow cover product using high resolution satellite observations. In Proceedings of the 32nd EARSeL Symposium 17, Mykonos Island, Greece, 21–24 May 2012.

60. Immitzer, M.; Vuolo, F.; Atzberger, C. First experience with Sentinel-2 data for crop and tree species classifications in central Europe. *Remote Sens.* **2016**, *8*, 166. [CrossRef]

61. Paul, F.; Winsvold, S.H.; Kääb, A.; Nagler, T.; Schwaizer, G. Glacier remote sensing using sentinel-2. Part II: Mapping glacier extents and surface facies, and comparison to Landsat 8. *Remote Sens.* **2016**, *8*, 575. [CrossRef]

62. Toming, K.; Kutser, T.; Laas, A.; Sepp, M.; Paavel, B.; Nõges, T. First experiences in mapping lake water quality parameters with Sentinel-2 MSI imagery. *Remote Sens.* **2016**, *8*, 640. [CrossRef]

63. Huang, H.; Roy, D.P.; Boschetti, L.; Zhang, H.K.; Yan, L.; Kumar, S.S.; Gomez-Dans, J.; Li, J. Separability analysis of Sentinel-2A multi-spectral instrument (MSI) data for burned area discrimination. *Remote Sens.* **2016**, *8*, 873. [CrossRef]

64. Lefebvre, A.; Sannier, C.; Corpetti, T. Monitoring urban areas with Sentinel-2A data: Application to the update of the Copernicus high resolution layer imperviousness degree. *Remote Sens.* **2016**, *8*, 606. [CrossRef]

65. Radoux, J.; Chomé, G.; Jacques, D.C.; Waldner, F.; Bellemans, N.; Matton, N.; Lamarche, C.; D'Andrimont, R.; Defourny, P. Sentinel-2's potential for sub-pixel landscape feature detection. *Remote Sens.* **2016**, *8*, 488. [CrossRef]

66. Drusch, M.; Del Bello, U.; Carlier, S.; Colin, O.; Fernandez, V.; Gascon, F.; Lamarche, C.; D'Andrimont, R.; Meygret, A. Sentinel-2: ESA's optical high-resolution mission for GMES operational services. *Remote Sens. Environ.* **2012**, *120*, 25–36. [CrossRef]

67. Malenovský, Z.; Rott, H.; Cihlar, J.; Schaepman, M.E.; García-Santos, G.; Fernandes, R.; Berger, M. Sentinels for science: Potential of Sentinel-1,-2, and-3 missions for scientific observations of ocean, cryosphere, and land. *Remote Sens. Environ.* **2012**, *120*, 91–101. [CrossRef]

68. Li, S.; Ganguly, S.; Dungan, J.L.; Wang, W.; Nemani, R.R. Sentinel-2 MSI radiometric characterization and cross-calibration with Landsat-8 OLI. *Adv. Remote Sens.* **2017**, *6*, 147–159. [CrossRef]

69. Zhu, Z.; Woodcock, C.E. Object-based cloud and cloud shadow detection in Landsat imagery. *Remote Sens. Environ.* **2012**, *118*, 83–94. [CrossRef]

70. Siljamo, N.; Hyvarinen, O.; Koskinen, J. Operational Snowcover Mapping using MSG/SEVIRI Data. In Proceedings of the IEEE International IGARSS Geoscience and Remote Sensing Symposium, Boston, MA, USA, 7–11 July 2008; Volume 5, p. V-45.

71. Julitta, T.; Cremonese, E.; Migliavacca, M.; Colombo, R.; Galvagno, M.; Siniscalco, C.; Rossini, M.; Fava, F.; Cogliati, S.; Morra di Cella, U.; et al. Using digital camera images to analyse snowmelt and phenology of a subalpine grassland. *Agric. For. Meteorol.* **2014**, *198*, 116–125. [CrossRef]

72. Hall, D.K.; Riggs, G.A.; Salomonson, V.V.; DiGirolamo, N.E.; Bayr, K.J. MODIS snow-cover products. *Remote Sens. Environ.* **2002**, *83*, 181–194. [CrossRef]

73. Hall, D.K.; Riggs, G.A.; Salomonson, V.V. Algorithm theoretical basis document (ATBD) for the MODIS snow and sea ice-mapping algorithms. 2001. Available online: https://eospso.gsfc.nasa.gov/sites/default/files/atbd/atbd_mod10.pdf (accessed on 1 December 2018).

74. Painter, T.H.; Rittger, K.; McKenzie, C.; Slaughter, P.; Davis, R.E.; Dozier, J. Retrieval of subpixel snow covered area, grain size, and albedo from MODIS. *Remote Sens. Environ.* **2009**, *113*, 868–879. [CrossRef]

75. Solberg, R.; Wangensteen, B.; Metsämäki, S.; Nagler, T.; Sandner, R.; Rott, H.; Pulliainen, J. *GlobSnow Snow Extent Product Guide Product Version 1.0*; European Space Angency: Helsinki, Finland, 2010.

76. Louis, J.; Debaecker, V.; Pflug, B.; Main-Korn, M.; Bieniarz, J.; Mueller-Wilm, U.; Cadau, E.; Gascon, F. Sentinel-2 Sen2Cor: L2A Processor for Users. *Living Planet Symp.* **2016**, *740*, 91.

77. Gao, B.C.; Goetz, A.F.H.; Westwater, E.R.; Conel, J.E.; Green, R.O. Possible near-IR channels for remote sensing precipitable water vapor from geostationary satellite platforms. *J. Appl. Meteorol.* **1993**, *32*, 1791–1801. [CrossRef]

78. Siljamo, N.; Hyvärinen, O. New Geostationary Satellite–Based Snow-Cover Algorithm. *J. Appl. Meteorol. Climatol.* **2011**, *50*, 1275–1290. [CrossRef]

79. Derrien, M.; Le Gléau, H. MSG/SEVIRI cloud mask and type from SAFNWC. *Int. J. Remote Sens.* **2005**, *26*, 4707–4732. [CrossRef]

80. Dybbroe, A.; Karlsson, K.G.; Thoss, A. NWCSAF AVHRR cloud detection and analysis using dynamic thresholds and radiative transfer modeling. Part I: Algorithm description. *J. Appl. Meteorol.* **2005**, *44*, 39–54. [CrossRef]

81. Kidder, S.Q.; Wu, H.T. Dramatic contrast between low clouds and snow cover if daytime 3.7 imagery. *Mon. Weather Rev.* **1984**, *112*, 2345–2346. [CrossRef]

82. Matson, M. NOAA satellite snow cover data. *Glob. Planet. Chang.* **1991**, *4*, 213–218. [CrossRef]

83. Derrien, M.; Le Gléau, H.; Fernandez, P. Algorithm Theoretical Basis Document for "Cloud Products" (CMa-PGE01 v3.2, CT-PGE02 v2.2 & CTTH-PGE03 v2.2). Available online: http://www.nwcsaf.org/AemetWebContents/ScientificDocumentation/Documentation/MSG/SAF-NWC-CDOP2-MFL-SCI-ATBD-01_v3.2.1.pdf (accessed on 1 December 2018).

84. Bunting, J.T.; d'Entremont, R.P. *Improved Cloud Detection Utilizing Defense Meteorological Satellite Program Near Infrared Measurements*; No. AFGL-TR-82-0027; Air Force Geophysics Laboratory: Hanscom AFB, MA, USA, 1982.

85. Dozier, J. Spectral signature of alpine snow cover from the Landsat Thematic Mapper. *Remote Sens. Environ.* **1989**, *28*, 9–22. [CrossRef]

86. Romanov, P.; Tarpley, D.; Gutman, G.; Carroll, T. Mapping and monitoring of the snow cover fraction over North America. *J. Geophys. Res. Atmos.* **2003**, *108*. [CrossRef]

87. Metsämäki, S.J.; Anttila, S.T.; Markus, H.J.; Vepsäläinen, J.M. A feasible method for fractional snow cover mapping in boreal zone based on a reflectance model. *Remote Sens. Environ.* **2005**, *95*, 77–95. [CrossRef]

88. Warren, S.G. Optical properties of snow. *Rev. Geophys.* **1982**, *20*, 67–89. [CrossRef]

89. Proy, C.; Tanre, D.; Deschamps, P.Y. Evaluation of topographic effects in remotely sensed data. *Remote Sens. Environ.* **1989**, *30*, 21–32. [CrossRef]

90. Smith, J.A.; Lin, T.L.; Ranson, K.J. The Lambertian assumption and Landsat data. *Photogramm. Eng. Remote Sens.* **1980**, *46*, 1183–1189.

91. Vikhamar, D.; Solberg, R.; Seidel, K. Reflectance modeling of snow-covered forests in hilly terrain. *Photogramm. Eng. Remote Sens.* **2004**, *70*, 1069–1079. [CrossRef]

92. Ertürk, A.G.; Barbosa, H. Detecting V-Storms using Meteosat Second Generation SEVIRI image and its applications: A case study over western Turkey. In Proceedings of the IEEE International Geoscience and Remote Sensing Symposium, IGARSS 2009, Cape Town, South Africa, 12–17 July 2009; Volume 3, p. III-609.

93. *WMO Guide to Meteorological Instruments and Methods of Observation*, 7th ed.; WMO-No. 8; WMO: Geneva, Switzerland, 2008.

94. Rittger, K.; Painter, T.H.; Dozier, J. Assessment of methods for mapping snow cover from MODIS. *Adv. Water Resour.* **2013**, *51*, 367–380. [CrossRef]

95. Metsämäki, S.; Vepsäläinen, J.; Pulliainen, J.; Sucksdorff, Y. Improved linear interpolation method for the estimation of snow-covered area from optical data. *Remote Sens. Environ.* **2002**, *82*, 64–78. [CrossRef]

96. Vikhamar, D.; Solberg, R. Snow-cover mapping in forests by constrained linear spectral unmixing of MODIS data. *Remote Sens. Environ.* **2003**, *88*, 309–323. [CrossRef]

97. Chang, A.; Foster, J.; Rango, A. The role of passive microwaves in characterizing snow cover in the Colorado River Basin. *GeoJournal* **1992**, *26*, 381–388. [CrossRef]

98. Hall, D.; Foster, J.; Chang, A. Measurement and modeling of microwave emission from forested snowfields in Michigan. *Hydrol. Res.* **1982**, *13*, 129–138. [CrossRef]

99. Klein, A.G.; Hall, D.K.; Riggs, G.A. Improving snow cover mapping in forests through the use of a canopy reflectance model. *Hydrol. Process.* **1998**, *12*, 1723–1744. [CrossRef]

100. Vikhamar, D.; Solberg, R. Subpixel mapping of snow cover in forests by optical remote sensing. *Remote Sens. Environ.* **2003**, *84*, 69–82. [CrossRef]

101. Painter, T.H.; Dozier, J.; Roberts, D.A.; Davis, R.E.; Green, R.O. Retrieval of subpixel snow-covered area and grain size from imaging spectrometer data. *Remote Sens. Environ.* **2003**, *85*, 64–77. [CrossRef]
102. Dietz, A.J.; Kuenzer, C.; Gessner, U.; Dech, S. Remote sensing of snow—A review of available methods. *Int. J. Remote Sens.* **2012**, *33*, 4094–4134. [CrossRef]
103. Andreadis, K.M.; Storck, P.; Lettenmaier, D.P. Modeling snow accumulation and ablation processes in forested environments. *Water Resour. Res.* **2009**, *45*. [CrossRef]
104. H-SAF: Product Validation Report for Product H10-SN-OBS-1. Available online: http://hsaf.meteoam.it/PVR-sn.php (accessed on 1 December 2018).
105. H-SAF: Product Validation Report for Product H12-SN-OBS-3. Available online: http://hsaf.meteoam.it/PVR-sn.php (accessed on 1 December 2018).

Article

Automated Classification of Terrestrial Images: The Contribution to the Remote Sensing of Snow Cover

Roberto Salzano [1,*], Rosamaria Salvatori [2], Mauro Valt [3], Gregory Giuliani [4,5], Bruno Chatenoux [5] and Luca Ioppi [6]

[1] National Research Council of Italy, Institute of Atmospheric Pollution Research, via Madonna del Piano 10, 50019 Sesto Fiorentino (FI), Italy

[2] National Research Council of Italy, Institute of Atmospheric Pollution Research, via Salaria km 29,300, 00015 Monterotondo (RM), Italy; salvatori@iia.cnr.it

[3] Veneto Regional Agency for Environmental Protection and Prevention, Arabba Avalanche Center, via Pradat 5, 32020 Arabba (BL), Italy; mauro.valt@arpa.veneto.it

[4] EnviroSPACE Lab, Institute for Environmental Sciences, University of Geneva, Bd Carl-Vogt 66, CH-1211 Geneva, Switzerland; gregory.giuliani@unige.ch

[5] Institute for Environmental Sciences, University of Geneva, GRID-Geneva, Bd Carl-Vogt 66, CH-1211 Geneva, Switzerland; bruno.chatenoux@unepgrid.ch

[6] Department of Science, University of Roma TRE, l.go San Leonardo Murialdo 1, 00146 Roma, Italy; luca.ioppi@uniroma3.it

* Correspondence: roberto.salzano@cnr.it; Tel.: +39-0555526587

Received: 11 December 2018; Accepted: 13 February 2019; Published: 19 February 2019

Abstract: The relation between the fraction of snow cover and the spectral behavior of the surface is a critical issue that must be approached in order to retrieve the snow cover extent from remotely sensed data. Ground-based cameras are an important source of datasets for the preparation of long time series concerning the snow cover. This study investigates the support provided by terrestrial photography for the estimation of a site-specific threshold to discriminate the snow cover. The case study is located in the Italian Alps (Falcade, Italy). The images taken over a ten-year period were analyzed using an automated snow-not-snow detection algorithm based on Spectral Similarity. The performance of the Spectral Similarity approach was initially investigated comparing the results with different supervised methods on a training dataset, and subsequently through automated procedures on the entire dataset. Finally, the integration with satellite snow products explored the opportunity offered by terrestrial photography for calibrating and validating satellite-based data over a decade.

Keywords: fractional snow cover; remote sensing; terrestrial photography; cold regions

---

1. Introduction

Snow cover is an important component of the cryosphere that plays a key role for climate dynamics and the resources availability: the seasonality of the snow cover influences, in fact, weather patterns, hydropower generation, agriculture, forestry, tourism, and aquatic ecosystems [1–3]. Remote sensing is the most common tool for the routine estimation of the snow cover extent. However, two different aspects must be considered for the enhancement of the final output: time and spatial resolutions. Both components, using remotely sensed data, are connected to each other, since the higher the spatial resolution (below hundreds of meters), the lower the revisit time interval (more than 1 week) [4].

The state-of-the-art snow products concerning the snow extent are remotely sensed and they are based mainly on multispectral optical sensors. They can investigate the snow cover and give information about the size and the shape of snow grains [5]; the presence of impurity soot; the age of the snow; and the presence of depth hoars. Furthermore, the short-wave infrared signal can support the discrimination between snow and clouds [6]. The estimation of the snow extent from remotely sensed multispectral images is based on the relation between the radiative behavior of the surface and the Fractional Snow Cover (FSC). This parameter describes the percentage of surface covered by snow [7] in a pixel element of a remotely sensed image. Considering that snow-covered surfaces are highly reflective in the visible range and low reflective in the short-wave infrared (swir) [8], it is possible to define an index that enhances the discrimination between snow and not snow in a single pixel. This index, defined as Normalized Difference Snow Index (NDSI), is calculated as follows:

$$\text{NDSI} = \frac{\text{green} - \text{swir}}{\text{green} + \text{swir}} \tag{1}$$

The green and the swir parameters are the bands available for each satellite sensor and their selection includes generally wavelength ranges centered at 500–600 nm (green) and 1500–1600 nm (swir). The relation between the FSC and the Normalized Difference Snow Index (NDSI) represents the most common inference required by remote sensing studies. There are two options for estimating the NDSI—FSC relation: the first one consists in combining satellite products with different spatial resolution [9,10]; and the second one can be approached having a ground truth information. The first solution is based on [8] combining Landsat and MODIS data and a NDSI to FSC relation is defined.

$$\text{FSC} = 1.45 \times \text{NDSI} - 0.01 \tag{2}$$

This knowledge is implemented in the SNOWMAP algorithm [11], which is the core of the MODIS data chain for the definition of remotely sensed snow products. The second solution can be approached defining an empirical reflectance-to-snow-cover model that requires a calibration having a number of reference sites in the satellite image. The most important example is the so-called Norwegian Linear Reflectance-to-snow-cover algorithm (NLR) [12] that is the core of the GlobSnow Snow Extent (SE) data chain [13]. From this perspective, the availability of webcam networks is an important data source for calibration and validation processes. The attention of the scientific community of this proxy is increasing, and the literature about this topic is growing [9,14–17]. Furthermore, several tools (for example, FMIPROT and PRACTISE) can be considered for research purposes [18–20]. The solutions available for the analysis of webcam imagery are commonly based on two different processes: orthorectification and classification. While the geometrical issue is based on the mathematical solution of the relationship between pixel elements and the ground surface, the detection of snow cover represents the real cognitive gap. The classification issue can be approached, following the applications available for the remote sensing imagery, using supervised, unsupervised or object-oriented methods [21], depending on the number of images that must be processed.

The focus of this paper is to investigate the contribution of the terrestrial photography to define site-specific thresholds useful for studying the snow cover with remotely sensed data. The expected outcomes are: (i) the description of an automated procedure able to process long time series of ground-based images; (ii) the comparison between automated approaches and supervised methods; (iii) and the evaluation of the potential contribution of terrestrial photography to the snow cover retrieval from remotely sensed data.

2. Methods

The purposes of this study required the investigation of different components and the integration of different data sources (Figure 1). The accomplishment of the declared objectives was approached selecting a study site where ground-based cameras were positioned for a decade. The first part of the effort was devoted to the analysis of the available terrestrial dataset. In this case, the selection of the most appropriate procedure was obtained considering the automated procedures and the supervised methods in order to check the overall performance of automated solutions under different conditions of illumination and snow coverage. Secondly, the collection of different satellite products provided the material useful for evaluating the potential impact of terrestrial photography on the estimation of snow extent from remotely sensed data.

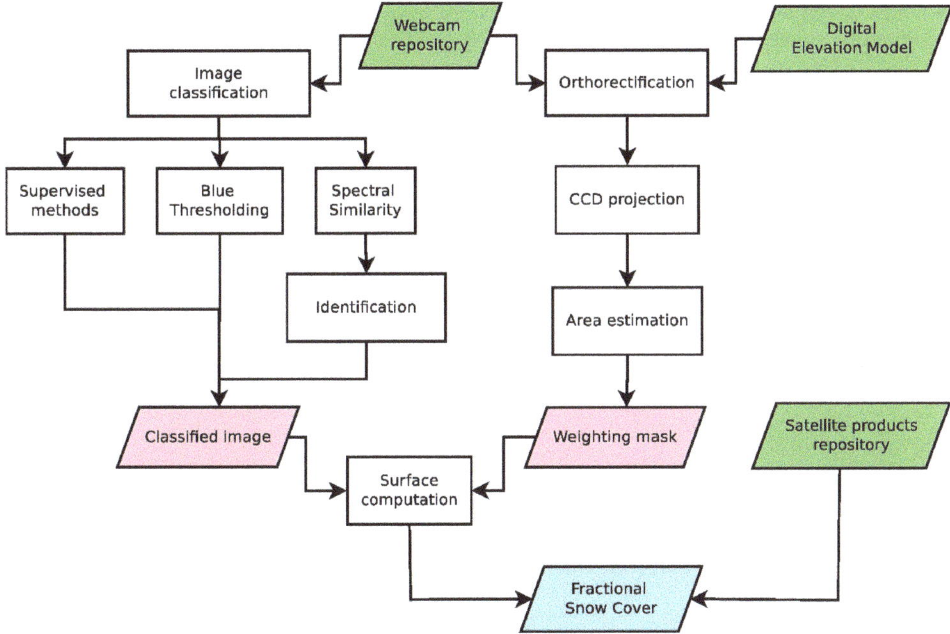

Figure 1. Description of the workflow followed in the manuscript. While the green boxes represent the considered data sources, the other colored boxes constitute the final products obtained by the different procedures required for the estimation of the Fractional Snow Cover.

2.1. Study Area

The considered study area (Figure 2) is located in the Italian northeastern Alps (Lago di Cavia, Falcade, Italy). The webcam (46°21′24″ N, 11°49′20″ E, WGS84) was positioned in a ski resort at 2200 m above the sea level. The study site is characterized by a snow cover duration almost complete from mid-November to late April, a melting season completed at the beginning of June and occasional snowfall in the rest of the year [22]. The selection of the site for the camera setup was supported by the topographic behavior of the location, which is an almost flat area with a soft slope where an artificial water body is located. The presence of an important ski facility and the management of this water resource outline the importance of this location.

2.2. Camera Setup

The webcam system was provided by Sistemi Video Monitoraggio S.r.l. (Romito Magra-SP) using a digital camera (Olympus C765). The camera was deployed at 2 m above the ground. The camera was featured by 4-megapixel sensor and a 1/2.5″ CCD, the focal length was 6.3 mm and images were saved in the jpeg data format with an 800 × 600 pixel resolution. Data logging and transmission were provided by specific hardware placed into a waterproof case and the power supply was ensured by the direct connection to the electric mains and by photovoltaic panels with a buffer battery. Data transfer was performed using an intranet connection with the receiving station located in Arabba through a mobile connection. The Veneto Regional Agency for Environmental Protection and Prevention developed a webcam section in the website (www.arpa.veneto.it) that supported the near-real-time availability of the images. The field of view defined by the camera perspective considered an area of about 5000 m² with a maximum distance from the camera up to 180 m. The camera acquired all-year-round images every 1 hour since 2004 to 2013. For this study, we considered a "complete" dataset with about 8000 images where every melting season was included in order to have a large range of snow cover and illuminating conditions. In addition of that, we defined a "small" dataset with 30 images dating back to 2008 and 2009, which included a large variability in terms of illuminating conditions and snow cover.

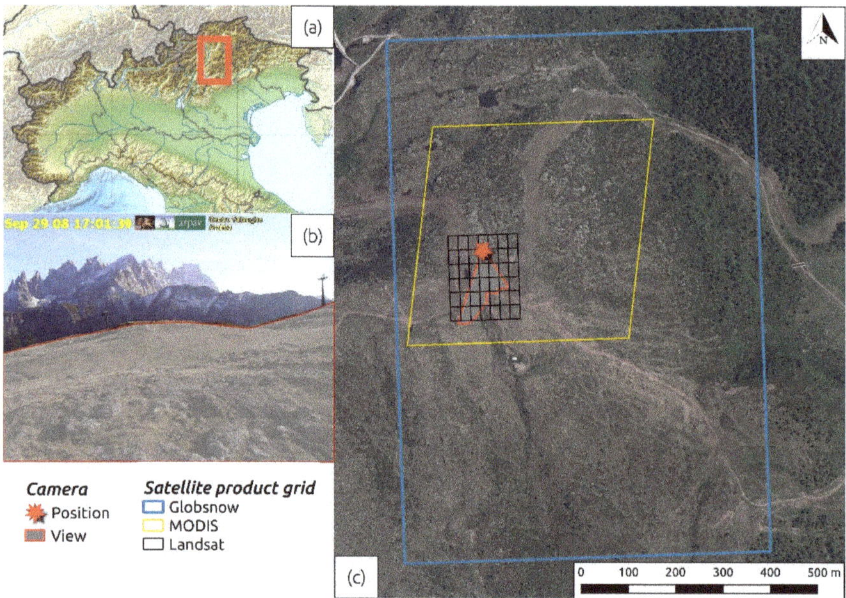

Figure 2. The study site of Cima Pradazzo (**a**), close to Falcade (Italy). Panoramic view of the camera (**b**) with the considered mask in red. The orthorectified views of the camera (**c**): the grey shaded area with red contour shows the camera view projected on the ground; and the colored lines indicate the pixel grids of the different satellite products.

2.3. Terrestrial Image Classification

Following the guidelines developed for the analysis of multispectral remotely sensed images, the classification issue can be approached using different principles depending on the methods for measuring the spectral matching or the spectral similarity: the deterministic-empirical methods and the stochastic approaches [23]. The deterministic measures include the spectral angle, the Euclidean distance and the cross-correlation of spectral vectors in the hyperspectral space. The stochastic measures evaluate the statistical distributions of spectral reflectance values of the targeted region of interests. Within this framework, a large variety of classification methods that can be grouped from different perspectives [24].

2.3.1. Supervised Methods

The requirement for the automated solution is a "parametric" method, based on a "per-pixel" classification about the presence of snow cover. The description of the pixel content must be definitive and, consequently, a "hard" classifier is necessary. Furthermore, the classification process cannot be iterative and specific for a single image. Consequently, the generalization for different images, under different illumination conditions, can be obtained with a "supervised" classifier, which considers a "training" Region Of Interest (ROIs) associated with the theoretical "white" snow. Looking at the supervised methods, we can include classifiers that are sensitive to the user experience during the definition of the region of interests and to the selection of discriminant parameters between snow and not snow. Some methods are associated with the threshold selection defined by the statistics of the identified ROIs. This is the case, for example, of the Parallelepiped classifier (PA), where the user defines a threshold based on the standard deviation. Some other approaches consider the probability associated with a specific ROI [25], calculating the Euclidean distance for the Minimum Distance (MD) method, the Mahalanobis distance for the Mahalanobis classifier (MA), and the covariance-based discriminant function for the Maximum Likelihood method (ML). These algorithms are all implemented in the commercial suite ENVI version 4.7 (Exelis Visual Information Solutions, Boulder, Colorado).

2.3.2. Blue Thresholding

Within the group of automated methods, there is a well-established method that is currently in use for snow-cover purposes with some limitations: it is a linear classifier based on thresholding of the blue channel (BT) that was introduced by [26] in the Snow-noSnow software. The method is based on the frequency counting of the blue component, and its hardness is associated with the definition of snow-not-snow limit looking at increments in the blue-channel histogram. This method has been used in several studies and it has shown some limitations. The illuminating conditions, the surface roughness and the distance from the camera are critical issues that affect the performance on retrieving snowed covers [27]. These limitations are the grounds of research for a higher performing method that possibly increases the depth of analyzing RGB imagery.

2.3.3. Spectral Similarity

The approach proposed in this paper is based on measuring the spectral variations in a 3D color space where reference endmembers are a theoretical "white" snow and a theoretical "black" target. The parameters estimated in this vector system are the spectral angle defined by [28] and the Euclidean distance [21], respectively calculated considering white and black references. While the parameter based on the Spectral Similarity (SS) represents an independent spectral feature, the Euclidean distance of the vector can be defined as a brightness-dependent feature. The involvement of all the three-color components will support the increase of surface types that can be discriminated: snow, shadowed snow and not snow. The proposed approach (Figure 1) was developed in the R programming environment [29].

The first step consists of rearranging the three-color components of each pixel into a new two-dimensional vector space, mathematically defined as follows:

$$\theta = acos\frac{P_R R_R + P_G R_G + P_B R_B}{\sqrt{P_R^2 + P_G^2 + P_B^2}\sqrt{R_R^2 + R_G^2 + R_B^2}} \tag{3}$$

The spectral angle θ in Equation (3) represents the relative proportion of the three-pixel components (P_R, P_G and P_B) in relationship to the reference composition ($R_R = 1$, $R_G = 1$ and $R_B = 1$). The angle varies from zero, which can be associated with a "flat" behavior of colors (R = G = B), to $\frac{\pi}{2}$, referring to a very dissimilar behavior from the theoretical "white" reference.

$$\Delta = \sqrt{P_R^2 + P_G^2 + P_B^2} \tag{4}$$

The spectral distance Δ in Equation (4) is conversely an estimation of the vector length in the RGB space. It can range from 0 (black) to 1.73 (white) and it can be associated with the Euclidean distance from a "black" reference RGB composition ($R_R = 0$, $R_G = 0$ and $R_B = 0$). While this parameter is sensitive to the brightness of colors, the spectral angle is invariant with brightness [23]. The outcome of this step consists in the frequency counting of pixels considering the two spectral components with a 0.05 resolution. Furthermore, the total number of included pixels (f_{tot}) and the area included in the cluster perimeter (P_f) were estimated.

The second step of the procedure consists in discriminating clusters from the obtained frequency distribution, and a watershed algorithm [30] can support this segmentation phase. Each cluster was fitted with a normal distribution in order to retrieve modes (defined by μ_Δ and μ_θ) and deviations (σ_Δ and σ_θ). If clusters are very close to each other, they can be combined in one larger group depending on their probability to be discriminated using the Mahalanobis distance. The criteria adopted for the definition of the cluster perimeter was based on the pixel frequency $f(\Delta\prime, \theta\prime)$ higher than the Poisson error of the adjacent pixel $f(\Delta, \theta)$ (Equation (5)).

$$f(\Delta\prime, \theta\prime) > \sqrt{f(\Delta, \theta)} \tag{5}$$

The procedure for the delimitation of the cluster perimeter was implemented using a per-pixel method following [31].

The final step consists in the identification of the surface type (snow, not snow and shadowed snow). This step was defined observing the frequency distributions of pixels in the defined spectral space (Figure 3). It was possible to detect that snow covers were generally characterized by higher θ angles and lower Δ values than not-snow covers. Snowed centroids (defined by μ_Δ and μ_θ) were generally positioned where angles were higher than 0.9 and distances were lower than 0.1.

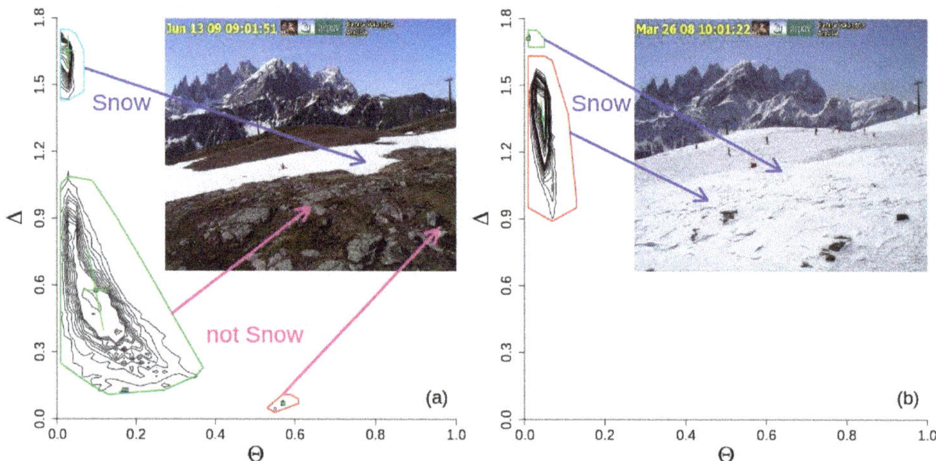

Figure 3. Examples of two different snow-not-snow mixtures. Colored polygons identified areas of clusters in presence of two different situations: partial (**a**) and full (**b**) snow cover. Lower plots are frequency distribution of pixels at the different spectral angles (θ) and spectral distances (Δ).

Furthermore, the range of cluster values (Δ_{max}, Δ_{min}, θ_{max} and θ_{min}) were characterized by short distance variations compared to angles in the case of snow-covered surfaces. From this point of view, clusters with limited perimeters ($P_f < 0.04$) and a high number of included pixels ($f_{tot} > 50$ of the analyzed pixels) described surfaces with homogeneous reflective behavior, as expected for snow-covered surfaces. The second rule that can be considered includes clusters with limited perimeters ($P_f < 0.04$) and consistent number of included pixels ($10 < f_{tot} < 50$ of the analyzed pixels). The optical behavior of those clusters must be coupled to their centroid position that must have low spectral angles ($\mu_\Delta < 0.5$). These constraints describe, also in this case, clusters characterized by a homogeneous spectral behavior coherent with a snow-covered surface. The third rule that completes the classification procedure consisted on estimating the range of Δ between the defined clusters in the image and on defining a threshold (T_Δ) that discriminates snow and other surface types. Two situations can occur for defining clusters above the threshold as snow-covered surface: one with multiple clusters (Equation (6)) and one with a single polygon (Equation (7)).

$$\mu_\Delta > max\Delta_{max} - min\Delta_{min} \tag{6}$$

$$\mu_\Delta > 0.8 \tag{7}$$

Once performed the classification, the amount of snow-covered surface was obtained adding the contribution of each cluster identified as snow covered. Furthermore, the quality of the final output was checked by the target area over the 10-year series of images. From this perspective, the ground control points were used to estimate eventually-induced shifting of the target view, and also to control the occurrence of adverse meteorological conditions (fog, clouds, intense raining/snowing) that could affect the image. Finally, the dataset was filtered from artifacts coupling this analysis to some basic tests about the file corruption and the image resolution.

2.4. Orthorectification

The orthorectification module was based on the geometrical correction of the perspective view. This step was implemented following [32]. The available digital elevation model [33], with a 5 m spatial resolution and 1 m vertical resolution, provided about 300 topographic points that were projected on the camera view (Figure 1c). The effectiveness of the correction was estimated considering eight ground control points.

2.5. Satellite Snow Products

Several satellite products are available for the remote sensing of the cryosphere and for this study we considered products obtained by optical sensors, characterized by different spatial resolutions: high (below 100 m); intermediate (below 1 km); and low (higher than 1 km). The integration between those products and ground-based imagery will be tested, in order to improve the dataset concerning the snow cover over a decade.

2.5.1. Optical Remote Sensing with High Spatial Resolution

The available remotely sensed snow products with a higher spatial resolution (below 100 m) were limited to Landsat missions, considering the studied time range (2004–2013). The selected sensors included Landsat satellites from 5 to 8, taking some differences into account in terms of band spectral ranges. All these data are now processed and available in the Swiss Data Cube [34]. The Landsat satellites are characterized by a spatial resolution of 30 m and a revisit time of 16 days. The considered data were geometrically and atmospherically corrected (Level 2A) using the Second Simulation of the Satellite Signal in the Solar Spectrum (6S) algorithm available in the Atmospheric and Radiometric Correction of Satellite Imagery (ARCSI) software [35]. The final estimation of NDSI was possible considering Eq.1 and the first short-wave infrared band of Landsat sensors. The wavelength ranges are specific for each sensor and they correspond to 520–600 nm and 1550–1755 nm for missions 5 and 7, and 533–590 nm and 1566–1651 nm for mission 8 [36].

2.5.2. Optical Remote Sensing with Intermediate Spatial Resolution

The highest time resolution available for optical remote sensing at our latitudes is provided, in the framework of the Earth Observing System (EOS) flagship, by NASA's satellites Terra and Aqua. Both platforms are equipped with the Moderate Resolution Imaging Spectroradiometer (MODIS) and they provide the coverage of the Earth two times daily (Terra in the local morning and Aqua in the local afternoon). The instrument is characterized by 36 bands with a spatial resolution of 250 m in the visible range and 500 m in the short-wave infrared. The NASA's data chain provides the retrieval of

NDSI at the ground, and we considered the MOD10A1_006 and the MYD10A1_006 products, for Terra and Aqua respectively, obtained using the National Snow and Ice Data Center services [11]. The NDSI values, calculated using the MODIS bands 4 (545–565 nm) and 6 (1628–1652 nm) in Equation1, were obtained in absence of clouds 2314 times over 6556 overpasses within the studied period.

2.5.3. Optical Remote Sensing with Low Spatial Resolution

The daily estimation of the snow cover extent is being provided, over the considered period, by the European Space Agency as a component of the Data User Element. The GlobSnow Snow Extent (SE) product covers the Northern Hemisphere and it is going to be extended to the Sentinel missions. The GlobSnow SE processing system applies optical measurements in the visual and in the thermal part of the electromagnetic spectrum acquired by the ERS-2 sensor ATSR-2 and the Envisat sensor AATSR. The first step of the data chain is based on a cloud-cover retrieval algorithm (SCDA) where clouds, as well as large water bodies (oceans, lakes and rivers) and glaciers, are masked out. This algorithm is based on the brightness - temperature difference between 11 and 3.7 μm and on a set of additional rules, useful for certain sky conditions. Furthermore, the snow cover information is retrieved for not-vegetated areas by the NLR algorithm [37] where the band 2 (670 nm) is considered. This step is based on a semi-empirical reflectance model, where reflectance from a target is expressed as a function of the snow fraction. The Fractional Snow Cover can then be derived from the observed reflectance based on the given reflectance constants and the transmissivity values. The product is provided daily with a spatial resolution of 1 km and the data are available using the GlobSnow service [13].

2.6. Statistical Analysis

The statistical analysis performed on the available datasets was carried out using state-of-the-art tools [38] that were implemented in the R-Project programming environment [29].

3. Results

Results will be presented separating the three objectives of the paper. The first part of the analysis will consider a small dataset where different supervised and automated classifiers will be compared. The second section will consider a ten-year dataset where about 8000 images will be processed using automated solutions. Finally, the FSC estimated by terrestrial photography will be compared to the output obtained by remotely sensed data.

3.1. Comparison between Supervised and Automated Classifiers

This first part of the analysis includes two steps: one dedicated to the orthorectification of the panoramic view observed by the webcam; the other focused on the image classification performed considering the color components associated with a RGB color space. The first process produced a weighting mask applying a geometrical correction and all the considered classification algorithms used this product successively. The classification step was operated on a small dataset of 30 images due to the user intervention required by the supervised methods (ML, MD, MA and PD) for the definition of snowed ROIs. This is a strong limitation for the analysis of long time series, and it outlines the need of automated solutions since BT and SS algorithms, for example, did not require any user decision. The results obtained by the BT method and the SS algorithm were preliminarily analyzed considering the confusion matrix of each image and estimating the average overall accuracy as reported in Table 1.

Considering only two classes of cover (snow and not snow pixels), the comparison between automated and supervised classifiers showed in general a good agreement with an overall accuracy higher than 90%. Furthermore, SS showed a better performance compared to BT with an increased average accuracy of about 1–2% in terms of pixel number. While BT reached the full agreement with the supervised methods in 10% of images, SS matched the classifications obtained by the traditional approaches in more than 30% of images. The goodness of the automated algorithms is confirmed by the Cohen's kappa coefficient, which increases from 0.89 for BT to 0.93 for SS. Both averages indicated very good agreements between supervised and automated solutions but they confirmed the increased performance of the algorithm based on Spectral Similarity. Although these differences may seem limited, the contribution of 2000–5000 pixels (in a masked part of the camera image of 250,000 pixels) in terms of surface can be important, depending on the distance of each pixel. The projection of each pixel on the surface could increase consistently from closer to faraway pixels. From this perspective, the impact of omissions and false discoveries on the projected area could be higher than the overall accuracy in terms of pixels and it should be analyzed case by case.

Table 1. Overall accuracy of automated algorithms, Blue Thresholding (BT) and Spectral Similarity (SS), versus supervised classifiers: the Mahalanobis distance (MA); the Maximum Likelihood (ML); the Minimum Distance (MD); and the Parallelepiped classifier (PD).

	Overall Accuracy (%)			
	MA	**ML**	**MD**	**PD**
BT	96.9	96.8	97.9	97.8
SS	97.9	98.5	99.2	98.6

3.2. Comparison between Automated Classifiers

The comparison between the estimated snow-covered areas obtained by the two automated algorithms (Figure 4a and Figure S1 for one example) confirmed the trend on underestimating the snow extent by BT compared to SS (see Table S1 for the raw data). The FSC estimated by the two methods differed slightly (the non-parametric Kruskal-Wallis chi-squared test indicated a non-significant statistical difference) and the Root Mean Squared Error (RMSE) was about 7.4%. The relation between the two FSC estimations showed a good correlation (R^2 close to 0.95) and the slope of the regression was 0.91 with an intercept of 11.5%.

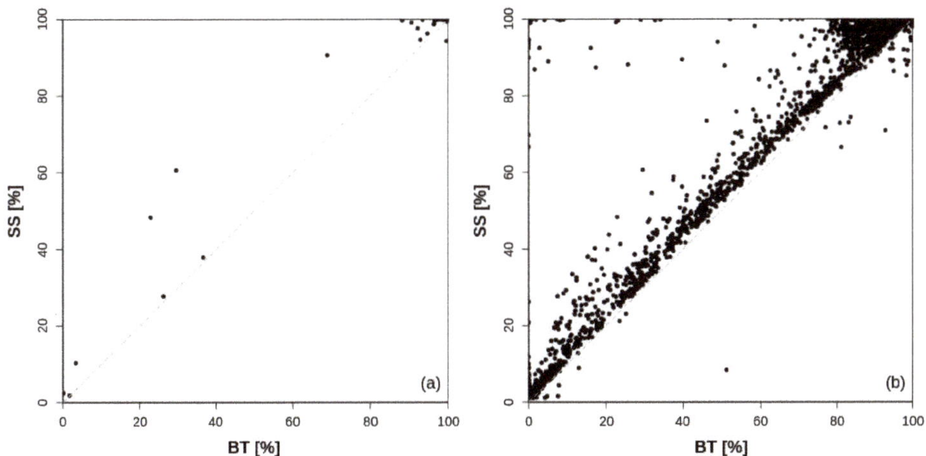

Figure 4. Performance of Blue Thresholding (BT) algorithm versus the Spectral Similarity (SS) method considering only the test dataset (**a**). Comparison between the two methods considering the complete dataset (**b**).

Although BT and SS estimations were almost consistent considering only the small dataset, the complete dataset highlighted an improved performance of SS (Figure 4b). The Kruskal-Wallis test indicated differences with a significance level higher than 99% and the RMSE was about 12%. The relation between the two FSC estimations showed a limited correlation (R^2 close to 0.87) compared to the small dataset, and the slope of the regression was 0.87 with an intercept of 14.5%. The detection of snow-covered areas using SS was generally higher than that obtained by BT and in few occurrences, it was completely missed by BT (see Table S2 for the raw data). The points closer to the left axis were, in fact, situations where light conditions (low sun elevation or intense cloud cover) affected the BT output. Those illumination conditions were important also in additional cases, where BT underestimated the snow-covered area compared to SS.

3.3. Comparison between FSC Estimations Obtained by Terrestrial Photography and Remote Sensing

The comparison between satellite products and terrestrial photography retrievals was focused on evaluating the relationship associated with the two data sources (see Table S3 for the raw data). We considered remotely sensed data with different spatial resolutions and data chains. The Landsat images available in the considered time range was 189, but 55 images were discarded due to the intense cloud coverage. The MODIS values were obtained in absence of clouds 2314 times over 6556 overpasses within the studied period. Finally, 289 GlobSnow data points were available during the considered period. While Landsat and MODIS data were converted in FSC considering the state-of-the-art relation described by [8], the GlobSnow product is ready-to-be-used considering the ground-truth support of the calibration sites identified in the images.

The Landsat sensors provided 24 observations (Figure 5a) and 10 were characterized by NDSI higher than 0.6, indicating the total coverage of snow in pixels. While two observations showed coherent NDSI values with the camera estimates (when snow cover was absent, the NDSI was negative), intermediate values were 3 times slightly above the expected results estimated using Equation (2) and 9 times consistently higher (more than 30% of overestimation). Whereas illumination differences can be related to the definition of a possible site-specific relation, heavy differences occurred when a partial shadow of clouds on the ground was present during the satellite revisit. The non-parametric Kruskal-Wallis chi-squared test indicated differences with a significance level of 80%, the RMSE was about 21% and the correlation coefficient was 0.59.

The MODIS sensors provided 430 observations (Figure 5b) and 205 were characterized by NDSI higher than 0.6, indicating the total coverage of snow in terms of pixels. The intermediate values were, also in this case, generally above the expected results. A first group of 26 observations showed camera FSC higher than expected NDSI-derived values with a difference higher than 30%; 33 observations were up to 30% higher; and 15 times MODIS products didn't detect any snow cover while the camera measured FSC ranging between 10–60%. All of these situations occurred when the cloud screening missed to identify partial cloud shadows on the ground while the satellite was overpassing. This comparison, in addition to Landsat indications, showed negative estimations in eight cases. These estimations (more than 20%) were artifacts associated with wrong cloud masking (there was no snow on the ground and it was full of clouds in the sky). The non-parametric Kruskal-Wallis chi-squared test indicated differences with a significance level of 99%, the RMSE was about 14% and the correlation coefficient was 0.91.

Finally, the GlobSnow SE product provided 62 observations (Figure 5c) and the estimated output was coherent 57 times (with full snow coverage at the ground), whereas the GlobSnow product missed to detect the snow cover 5 times, compared to the camera observations. The non-parametric Kruskal-Wallis chi-squared test indicated differences with a significance level of 99%, the RMSE was about 18% and the correlation coefficient was 0.84. From a statistical point of view, all the satellite products showed significant differences compared to the camera-based estimations even if the correlation was good. This observation is influenced, of course, by the number of outliers included in the available dataset composed by the different satellite revisits, which depends mostly on cloud screening.

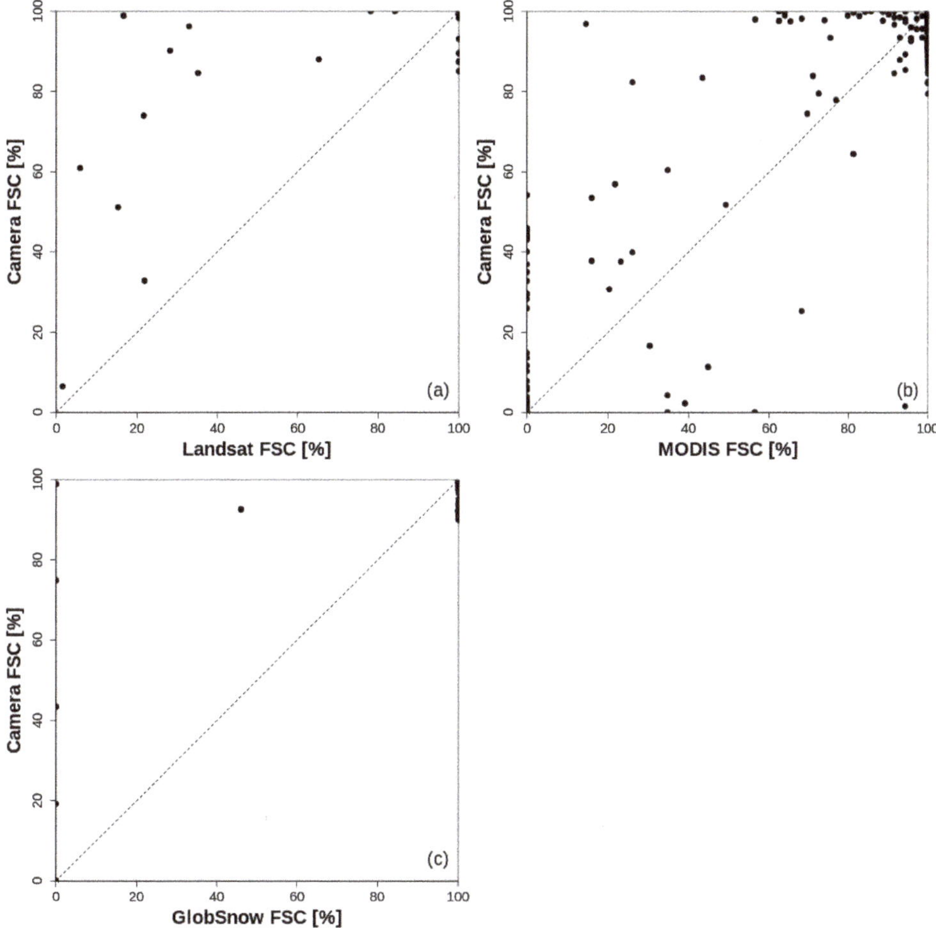

Figure 5. Comparison between Fractional Snow Cover estimations obtained by terrestrial photography and remote sensing. Plots refer to Landsat (**a**), MODIS (**b**) and GlobSnow (**c**).

4. Discussion

The first part of the results evidenced that automated solutions provide FSC estimations compatible to the supervised solutions available in literature. The major advantage of automated methods consists in the reduction of time consumption and, consequently, in the opportunity of processing long time series of terrestrial images. We described an automated approach based on the concept of Spectral Similarity [23], which could prevent artifacts under particular illuminating conditions. While a small training dataset supported the training of an SS-based algorithm, the ten-year dataset, with about 8000 images, showed a better performance compared to a state-of-the-art automated method BT described by [26]. The trend of FSC underestimation (about 10%) outlined by the small training dataset was confirmed by the large decadal comparison. The observed statistically significant differences were limited in terms of pixel number (less than 1%), but these discrepancies were important in terms of surface. The projection of each pixel on the surface could increase consistently from closer to faraway pixels and from this perspective; the impact of omissions and false discoveries on the projected area could be high. Furthermore, the ability to analyze the "difficult" conditions affecting the BT performance [10] was confirmed by statistically significant differences detected between the two data series. The limitations of BT retrievals can be associated with poor illuminating conditions (low Sun elevation or heavy cloud coverage) and surface roughness. While low Sun elevation can occur in the early morning or in the late afternoon, surface roughness and cloud coverage are not time dependent. Furthermore, while the illuminating conditions can alter the reflective behavior of snow in response to a more blueish incident light, the roughness can imply the presence of shadowed surfaces that BT cannot discriminate compared to SS. While BT tends to separate shadowed and illuminated areas, SS can be trained to integrate both types since the spectral angle is similar and its only variation is the spectral distance. While BT can generally provide good results between 11:00 am and 3:00 pm local time, SS can enlarge the range of performing conditions in terms of both Sun elevation and cloud cover. These preliminary results concerning the SS approach represent a first step towards the development of a machine learning strategy aimed to analyze routinely ground-based images. Artifacts associated with purely-BT classification [19,26,28], which are well documented in literature [19,27,28], were reduced and the need to consider all the information present in a RGB composite image [27] was followed. Differently from [27], which combined principal component analysis to BT, SS is independent from BT and considers all the bands at the beginning of the classification step obtaining a discrimination between surface types based directly on the spectral behavior of each classified feature. Furthermore, SS considers the color variations induced by illumination conditions and the probability to separate different surface types is associated with statistical measurements such as the Mahalanobis distance.

Finally, the FSC estimated by terrestrial photography and satellite products evidenced different aspects to be considered: the spatial resolution and the cloud screening. The cloud screening is a critical step present in all of the data chain considered in this study. Our data demonstrated, in fact, that a large number of satellite omissions were associated with a wrong detection of clouds. In addition to those exclusions, different situations evidenced an underestimation of FSC affected by the presence of cloud shadows that reduce the reflection of light from the surface. Although the different data chains [6,8,13] of course, consider these anomalies, the contribution of terrestrial photography, in this case, could support for the validation of remotely sensed retrievals. Moving to the spatial resolution, we considered data ranging from a 30 m resolution (Landsat), to 500 m (MODIS), to 1 km (GlobSnow SE) in order to test different data chains with different spatial and time resolutions. The spatial resolution had, of course, an impact and we found a more reliable relation with Landsat data than with those characterized by a coarser resolution. While the projected area of the camera view is five times the surface covered by a single Landsat pixel, it represents the 2% of a MODIS pixel the 0.5% of a single GlobSnow grid element. This implies that the surface morphology can affect the final estimates due to the presence of hills and small valleys.

This framework outlines the potentiality of facilities where different satellite snow products can have a common term of comparison such as terrestrial cameras. Ground-based images represent a good proxy, useful for validating the coherence between different products. On the one hand, this data-source can support the reconstruction of long time series useful for climate change studies. On the other one, this kind of proxy can assist the definition of site-specific relation between FSC and the optical behavior of the surface.

5. Conclusions

The contribution of terrestrial photography for the definition of the relation between the Fractional Snow Cover and the spectral behavior of the surface is a major issue. Ground-based cameras represent a valuable proxy of data useful for investigating the snow cover extension over a long period. From this perspective, terrestrial photography can be used as ancillary information and it supports the integration among different multispectral remotely sensed datasets. The availability of an automated procedure useful for the discrimination between snow and not-snow covered surfaces can support the analysis of large datasets. The selected approach based on Spectral Similarity was compared with supervised methods and with the Blue Thresholding procedure on a training dataset. Considering the supervised methods as a reference, the Spectral Similarity approach showed better performance in estimating the snow cover area. Furthermore, expanding the dataset to a 10-year terrestrial image record, the algorithm increased the capability to estimate the Fractional Snow Cover under a larger range of conditions compared to the state-of-the-art method. The integration with three different satellite snow products (Landsat, MODIS and GlobSnow) highlighted the potentiality to define a site-specific relation and threshold useful for isolating the snow cover area from remotely sensed data. Finally, the support provided by terrestrial photography enhanced the possibility to detect artifacts associated with clouds and shadows.

Supplementary Materials: The following are available online at http://www.mdpi.com/2076-3263/9/2/97/s1, Figure S1: Example of comparison between output obtained by automated classifications on a terrestrial image, Table S1: Training dataset composed by FSC estimated using supervised and automated methods (Figure 4a), Table S2: Complete dataset composed by FSC estimated using only automated methods (Figure 4b), Table S3: Complete dataset composed by FSC estimated using satellite products (Figure 5).

Author Contributions: R.S. (Roberto Salzano) and R.S. (Rosamaria Salvatori) worked on the algorithm concept, performed the analysis and prepared the manuscript; M.V. supported the definition of the image dataset, the data analysis and the interpretation of results; G.G. and B.C. provided the access to the Swiss Data Cube consequently making remotely sensed data available; L.I. performed the classification of images using supervised methods.

Acknowledgments: This work has been carried out within the ESSEM COST Action ES1404 and the authors would like to acknowledge the European Commission "Horizon 2020 Program" that founded the ERA-PLANET/GEOEssential project (Grant Agreement no. 689443). English revision and spell-check by Lena Rettori. We wish to acknowledge three anonymous reviewers that really helped to improve the quality of the paper.

Conflicts of Interest: The authors declare no conflict of interest.

References

1. Beniston, M.; Farinotti, D.; Stoffel, M.; Andreassen, L.M.; Coppola, E.; Eckert, N.; Fantini, A.; Giacona, F.; Hauck, C.; Huss, M.; et al. The European mountain cryosphere: A review of its current state, trends, and future challenges. *Cryosphere* **2018**, *12*, 759–794. [CrossRef]

2. Helmert, J.; Şensoy Şorman, A.; Alvarado Montero, R.; De Michele, C.; de Rosnay, P.; Dumont, M.; Finger, D.; Lange, M.; Picard, G.; Potopová, V.; et al. Review of Snow Data Assimilation Methods for Hydrological, Land Surface, Meteorological and Climate Models: Results from a COST HarmoSnow Survey. *Geosciences* **2018**, *8*, 489. [CrossRef]

3. Pirazzini, R.; Leppänen, L.; Picard, G.; Lopez-Moreno, J.I.; Marty, C.; Macelloni, G.; Kontu, A.; von Lerber, A.; Tanis, C.M.; Schneebeli, M.; et al. European In-Situ Snow Measurements: Practices and Purposes. *Sensors* **2018**, *18*, 2016. [CrossRef]

4. Dietz, A.J.; Kuenzer, C.; Gessner, U.; Dech, S. Remote sensing of snow–A review of available methods. *Int. J. Remote Sens.* **2011**, *33*, 4094–4134. [CrossRef]

5. Dozier, J.; Green, R.O.; Nolin, A.W.; Painter, T.H. Interpretation of snow properties from imaging spectrometry. *Remote Sens. Environ.* **2009**, *113*, 525–537. [CrossRef]

6. Rodell, M.; Houser, P.R. Updating a land surface model with MODIS-derived snow cover. *J. Hydrometeorol.* **2004**, *5*, 1064–1075. [CrossRef]

7. Painter, T.H.; Rittger, K.; McKenzie, C.; Slaughter, P.; Davis, R.E.; Dozier, J. Retrieval of subpixel snow-covered area and grain size from imaging spectrometer data. *Remote Sens. Environ.* **2009**, *113*, 868–879. [CrossRef]

8. Salomonson, V.V.; Appel, I. Development of the Aqua MODIS NDSI fractional snow cover algorithm and validation results. *IEEE Trans. Geosci. Remote* **2006**, *44*, 1747–1756. [CrossRef]

9. Yin, D.; Cao, X.; Chen, X.; Shao, Y.; Chen, J. Comparison of automatic thresholding methods for snow-cover mapping using Landsat TM imagery. *Int. J. Remote Sens.* **2013**, *34*, 6529–6538. [CrossRef]

10. Härer, S.; Bernhardt, M.; Siebers, M.; Schulz, K. On the need for a time- and location-dependent estimation of the NDSI threshold value for reducing existing uncertainties in snow cover maps at different scales. *Cryosphere* **2018**, *12*, 1629–1642. [CrossRef]

11. Hall, D.K.; Riggs, G.A. MODIS/[Terra/Aqua] Snow Cover Daily L3 Global 500m Grid, Version 6. NASA National Snow and Ice Data Center Distributed Active Archive Center. Available online: https://nsidc.org/data/modis/data_summaries (accessed on 1 October 2018).

12. Solberg, R.; Amlien, J.; Koren, H. A Review of Optical Snow Cover Algorithms. Norwegian Computing Center Note, SAMBA/40/06. 2006. Available online: https://www.nr.no/directdownload/4400/Solberg_-_A_review_of_optical_snow_algorithms.pdf (accessed on 1 October 2018).

13. Metsämäki, S.; Pulliainen, J.; Salminen, M.; Luojus, K.; Wiesmann, A.; Solberg, R.; Böttcher, K.; Hiltunen, M.; Ripper, E. Introduction to GlobSnow Snow Extent products with considerations for accuracy assessment. *Remote Sens. Environ.* **2015**, *156*, 96–108. [CrossRef]

14. Arslan, A.N.; Tanis, C.M.; Metsämäki, S.; Aurela, M.; Böttcher, K.; Linkosalmi, M.; Peltoniemi, M. Automated Webcam Monitoring of Fractional Snow Cover in Northern Boreal Conditions. *Geosciences* **2017**, *7*, 55. [CrossRef]

15. Kepski, D.; Luks, B.; Migała, K.; Wawrzyniak, T.; Westermann, S.; Wojtuń, B. Terrestrial Remote Sensing of Snowmelt in a Diverse High-Arctic Tundra Environment Using Time-Lapse Imagery. *Remote Sens.* **2017**, *9*, 733. [CrossRef]

16. Bradley, E.S.; Clarke, K.C. Outdoor Webcams as Geospatial Sensor Networks: Challenges, Issues and Opportunities. *Cartogr. Geogr. Inf. Sci.* **2011**, *38*, 3–19. [CrossRef]

17. Fedorov, R.; Camerada, A.; Fraternali, P.; Tagliasacchi, M. Estimating snow cover from publicly available images. *IEEE Trans. Multimed.* **2016**, *18*, 1187–1200. [CrossRef]

18. Parajka, J.; Haas, P.; Kirnbauer, R.; Jansa, J.R.; Blöschl, G. Potential of time-lapse photography of snow for hydrological purposes at the small catchment scale. *Hydrol. Process.* **2012**, *26*, 3327–3337. [CrossRef]

19. Härer, S.; Bernhardt, M.; Corripio, J.G.; Schulz, K. PRACTISE–Photo Rectification and ClassificaTIon SoftwarE (V.1.0). *Geosci. Model Dev.* **2013**, *6*, 837–848. [CrossRef]

20. Tanis, C.M.; Peltoniemi, M.; Linkosalmi, M.; Aurela, M.; Böttcher, K.; Manninen, T.; Arslan, A.N. A system for Acquisition, Processing and Visualization of Image Time Series from Multiple Camera Networks. *Data* **2018**, *3*, 23. [CrossRef]

21. Jensen, J.R. *Introductory Digital Image Processing: A Remote Sensing Perspective*, 4th ed.; Pearson Series in Geographic Information Science; Pearson: Glenview, IL, USA, 2015; pp. 1–544. ISBN 013405816X.

22. Valt, M.; Cianfarra, P. Recent snow cover variability in the Italian Alps. *Cold Reg. Sci. Tech.* **2010**, *64*, 146–157. [CrossRef]

23. Van der Meer, F. The effectiveness of spectral similarity measures for the analysis of hyperspectral imagery. *Int. J. Appl. Earth Obs. Geoinf.* **2006**, *8*, 3–17. [CrossRef]

24. Lu, D.; Weng, Q. A survey of image classification methods and techniques for improving classification performance. *Int. J. Remote Sens.* **2007**, *28*, 823–870. [CrossRef]

25. Richards, J.A.; Jia, X. *Remote Sensing Digital Image Analysis: An Introduction*, 4th ed.; Springer: Berlin, Germany, 2005; pp. 1–494. ISBN 3540251286.

26. Salvatori, R.; Plini, P.; Giusto, M.; Valt, M.; Salzano, R.; Montagnoli, M.; Cagnati, A.; Crepaz, G.; Sigismondi, D. Snow cover monitoring with images from digital camera systems. *Ital. J. Remote Sens.* **2011**, *43*, 137–145. [CrossRef]

27. Härer, S.; Bernhardt, M.; Schulz, K. PRACTISE–Photo Rectification and ClassificaTIon SoftwarE (V.2.1). *Geosci. Model Dev.* **2016**, *9*, 307–321. [CrossRef]

28. Kruse, F.A.; Lefkoff, A.B.; Boardman, J.W.; Heidebrecht, K.B.; Shapiro, A.T.; Barloon, P.J.; Goetz, A.F.H. The spectral image processing system (SIPS)-Interactive visualization and analysis of imaging spectrometer data. *Remote Sens. Environ.* **1993**, *44*, 145–163. [CrossRef]

29. R Core Team. *R: A Language and Environment for Statistical Computing*; ver. 3.4.4; R Foundation for Statistical Computing: Vienna, Austria, 2018; Available online: https://www.R-project.org/ (accessed on 1 October 2018).

30. Vincent, L.; Soille, P. Watersheds in Digital Spaces: An Efficient Algorithm Based on Immersion Simulations. *IEEE Trans. Pattern Anal. Mach. Intell.* **1991**, *13*, 583–598. [CrossRef]

31. Seo, J.; Chae, S.; Shim, J.; Kim, D.; Cheong, C.; Han, T. Fast Contour-Tracing Algorithm Based on a Pixel-Following Method for Image Sensors. *Sensors* **2016**, *16*, 353. [CrossRef] [PubMed]

32. Corripio, J.G. Snow surface albedo estimation using terrestrial photography. *Int. J. Remote Sens.* **2004**, *25*, 5705–5729. [CrossRef]

33. Regione del Veneto. Modello Digitale del Terreno dell'Intero Territorio Regionale con Celle di 5 Metri di Lato. 2006. Available online: http://idt.regione.veneto.it/app/metacatalogID:C0103024_DTM5 (accessed on 1 October 2018).

34. Giuliani, G.; Chatenoux, B.; De Bono, A.; Rodila, D.; Richard, J.; Allenbach, K.; Dao, H.; Peduzzi, P. Building an Earth Observations Data Cube: Lessons learned from the Swiss Data Cube (SDC) on generating Analysis Ready Data (ARD). *Big Earth Data* **2017**, *1*, 100–117. [CrossRef]
35. Vermote, E.F.; Tanre, D.; Deuze, J.L.; Herman, M.; Morcette, J.J. Second simulation of the satellite signal in the solar spectrum, 6S: An overview. *IEEE Trans. Geosci. Remote* **1997**, *35*, 675–686. [CrossRef]
36. Barsi, J.A.; Lee, K.; Kvaran, G.; Markham, B.L.; Pedelty, J.A. The Spectral Response of the Landsat-8 Operational Land Imager. *Remote Sens.* **2014**, *6*, 10232–10251. [CrossRef]
37. Metsämäki, S.; Mattila, O.P.; Pulliainen, J.; Niemi, K.; Luojus, K.; Böttcher, K. An optical reflectance model-based method for fractional snow cover mapping applicable to continental scale. *Remote Sens. Environ.* **2012**, *123*, 508–521. [CrossRef]
38. Congalton, R.G.; Green, K. *Assessing the Accuracy of Remotely Sensed Data: Principles and Practices*, 2nd ed.; CRC Press: Boca Raton, FL, USA, 2008; pp. 1–177. ISBN 978142005.

 *geosciences* 

Article

Multi-Source Based Spatio-Temporal Distribution of Snow in a Semi-Arid Headwater Catchment of Northern Mongolia

Munkhdavaa Munkhjargal [1,*], Simon Groos [2], Caleb G. Pan [3], Gansukh Yadamsuren [4], Jambaljav Yamkin [4] and Lucas Menzel [1]

[1] Professorship in Hydrology and Climatology, Department of Geography, Heidelberg University, Im Neuenheimer Feld 348, 69120 Heidelberg, Germany; lucas.menzel@uni-heidelberg.de

[2] Faculty of Spatial Information, Dresden University of Applied Sciences, Friedrich-List-Platz 1, 01069 Dresden, Germany; Simon.Groos@hotmail.de

[3] Numerical Terradynamic Simulation Group, University of Montana, Missoula, MT 59812, USA; Caleb.Pan@mso.umt.edu

[4] Permafrost Laboratory, Institute of Geography and Geoecology, Mongolian Academy of Science, Erkhuu Street, Sukhbaatar District, Ulaanbaatar UB-14192, Mongolia; gansukh.khatan@gmail.com (G.Y.); jambaljav@gmail.com (J.Y.)

* Correspondence: munkhjargal@uni-heidelberg.de; Tel.: +49-6221-54-5571

Received: 6 December 2018; Accepted: 14 January 2019; Published: 19 January 2019

Abstract: Knowledge of the duration and distribution of seasonal snow cover is important for understanding the hydrologic regime in mountainous regions within semi-arid climates. In the headwater of the semi-arid Sugnugur catchment (in the Khentii Mountains, northern Mongolia), a spatial analysis of seasonal snow cover duration (SCD) was performed on a 30 m spatial resolution by integrating the spatial resolution of Landsat-7, Landsat-8, and Sentinel-2A images with the daily temporal resolution of Moderate Resolution Imaging Spectroradiometer (MODIS) snow products (2000–2017). Validation was achieved using in situ time series measurements from winter field campaigns and distributed surface temperature loggers. We found a mean increase of SCD with altitude at approximately +6 days/100 m. However, we found no altitude-dependent changes in snow depth during field campaigns. The southern exposed valley slopes are either snow free or covered by intermittent snow throughout the winter months due to high sublimation rates and prevailing wind. The estimated mean SCD ranges from 124 days in the lower parts of the catchment to 226 days on the mountain peaks, with a mean underestimation of 12–13 days. Snow onset and melt dates exhibited large inter-annual variability, but no significant trend in the seasonal SCD was evident. This method can be applied to high-resolution snow mapping in similar mountainous regions.

Keywords: snow; snow cover duration; persistent and intermittent snow; optical remote sensing; northern Mongolia

---

1. Introduction

The duration and distribution of snow cover is an important governor of hydrologic and ecologic processes and contributes a significant role in local and regional hydrologic regimes [1–3]. For basin headwaters in arid or semi-arid regions, seasonal snow-melt sustains river discharge, contributing to regional water resources during dry summer periods. Hence, accurate estimation of snow cover area (SCA) is needed to evaluate the runoff from snowmelt. Snow cover also plays an important role in permafrost distribution through its insulating effect on seasonal ground temperature [4]. In response to recent accelerated warming, there has been an observed decrease in SCA and snow cover duration (SCD) in the Arctic regions [5]. In the mid-latitudes of Central Asia, snow cover has been observed to accumulate later into the autumn season [6] whilst the spring melt period coincides with increasing spring rainfall [7], reflecting mid-latitude extreme weather in winter months, coinciding with the arctic amplification and the strengthening of the Siberian high (SH) [8,9]. The SH creates an extreme continental climate that promotes low temperatures and longer seasonal snow cover extent, and is centrally located over northern Mongolia [10]. Our understanding of the timing and characteristics of snow cover is highly limited, yet it is critical, as Mongolia's climate is semi-arid, largely governed by the SH.

Snow cover can be accurately monitored using in situ measurements, including snow depth, density, and snow water equivalent (SWE); however, such measurements are limited for broad spatial analyses due to their sparse monitoring networks. In contrast, optical remote sensing applications are capable of detecting SCA globally with frequent temporal coverage but are often limited by cloud cover and winter time solar darkness.

Optical remote sensing data provide one of the best opportunities to characterize SCA and SCD in the data scarce region of Mongolia due to a high number of clear sky days during winter, promoted by the SH. The Moderate Resolution Imaging Spectroradiometer (MODIS) has provided daily snow cover products with 500 m spatial resolution since 2000, including MODIS Terra (MOD10A1) and Aqua (MYD10A1) [11–17]. In addition to MODIS, other popular optical sensors exist, including the Landsat series and more recently, the Sentinel-2A series, providing a higher spatial resolution but at a reduced temporal resolution [18], relative to MODIS. The combination of MODIS, Landsat, and Sentinel images [19] have proven to be useful to retrieve SCA by applying different downscaling techniques using air temperature, solar radiation, and topography [20–22]. However, the effectiveness in detecting SCA and SCD by such optical sensors is challenged by landscape heterogeneity created by topography, patchy snow cover, forest canopy, cloud coverage, and the spatial and temporal resolutions [13,23].

In this study, we combine the daily temporal resolution of MODIS Terra and Aqua with the high spatial resolution of Landsat and Sentinel images to derive SCA and SCD at a 30 m spatial resolution and at a daily temporal resolution over the Sugnugur catchment in northern Mongolia for winters 2000/2001 to 2016/2017. The catchment area (495 km^2) is located about 100 km northwest of Ulaanbaatar in northern Mongolia (Figure 1), with elevations ranging from 960 to 2800 m.a.s.l., forming the headwaters of the Selenga River before it drains into Lake Baikal. The importance of this small headwater catchment is highlighted for its snow-melt water contribution to regional water resources and energy exchange balance. Approximately 30% of the annual precipitation occurs as snowfall during the winter months [24,25], and 20% of annual evapotranspiration comes from snow sublimation and snow-melt water [26,27]. North-west prevailing winds are dominant and bring the majority of the precipitation, with a strong orographic effect. The lower elevations are defined by shrubs with short grasslands, whilst the Siberian boreal forest is found on north-facing slopes and higher elevations; yet about one third of this forest cover has been affected by wildfires [25].

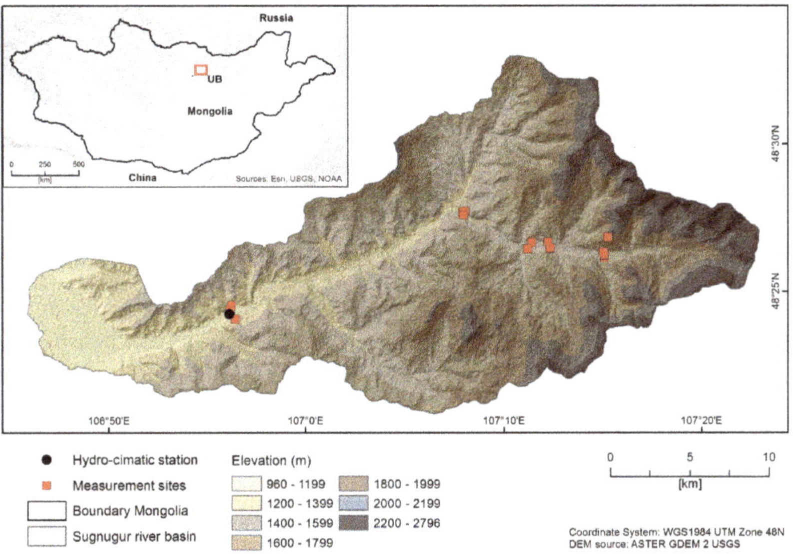

Figure 1. Sugnugur river catchment. Red points indicate the locations of snow field campaigns, as well as the surface temperature measurements. The hydro-climatic station is located at elevation 1193 m.a.s.l. UB is Ulaanbaatar.

We quantified the spatial distribution and timing of seasonal snow cover (i), introduced an alternative correction method to account for elevation and land cover when detecting SCA and SCD (ii), and reconstructed the snow cover development over the winter 2016/2017 (iii). We validated our results using several in situ observations including surface temperature, snow field measurements, and time-lapse photographs.

2. Materials and Methods

2.1. Materials

We used daily MODIS Terra (MOD10A1 V.006) and Aqua (MYD10A1 V.006) snow products (i), a digital elevation model (DEM) with 30 m spatial resolution (ii), Landsat-7, Landsat-8, and Sentinel-2A data (iii), a Landsat based land-cover classification map (iv), hydro-climatic data observed with high frequency (v), iButton surface temperature measurements distributed in the catchment (vi), and photographs from a time-lapse digital camera (vii). The datasets i–iv were used for reconstructing SCA and SCD whereas v–vii were used to validate the results.

MODIS snow-cover data: Daily MODIS snow products (Version. 006) from the Aqua and Terra satellites with 500 m spatial resolution were employed for the winters 2000/2001 to 2016/2017. These products contain normalized difference snow index (NDSI) values 0–100 as percentages, as well as other parameters, including cloud coverage. The MODIS snow products including the 8-day composite snow product are also widely used in many regions of the world [28–38]. However, the main drawbacks of these products are cloud coverage and coarse spatial resolution [39]. Therefore, we used the original daily data rather than using 8-day snow product to minimize the uncertainties that may have resulted from any cloud-filtering [15]. The data were obtained from the National Aeronautics and Space Administration (NASA) Earth Observing System Data and Information System (EOSDIS) Reverb platform [40].

Digital Elevation Model: We used a 30 m DEM derived from the Advanced Spaceborne Thermal Emission and Reflection Radiometer (ASTER) [41]. The DEM was used to derive the altitude-dependent rate of SCD to aggregate MODIS data to 30 m resolution.

Landsat and Sentinel data: Optical remote sensing data from the Landsat-7, Landsat-8, and Sentinel-2A were used to generate the winter snow cover time series [42]. The Landsat-7/8 data have a spatial resolution of 30 m and a temporal resolution of 16 days (8 days when they are combined). The Sentinel-2 has a spatial resolution of 20 m and a temporal resolution of 5 days. In total, 40 clear-sky day images (Table S1, Supplementary Materials) between the beginning of October 2016 (mid-autumn) and end of May 2017 (late spring) were processed. The Sentinel images were aggregated to 30 m resolution to meet the same Landsat and DEM resolutions.

Land-cover: Land-cover classification for Sugnugur catchment was produced based on a Landsat-7 image of the year 2011. We believed that this map was appropriate because there was no significant change in land-cover since 2011. In total, five different land-covers were classified in the Sugnugur catchment, including: forest, burned forest, short grassland, shrub, and rock. Areas with forest and shrubs were used in this study for correcting snow distribution maps from the Landsat and Sentinel images because of the low snow detection rate under canopy closure.

In-situ measurements: Snow depth (SD) has been observed using the Campbell SR50A-L sensor since winter 2012/2013 in short grassland vegetation, as well as other climatic parameters such as precipitation, air temperature, surface temperature, and albedo at our hydro-climatic station site located near the entry of the catchment. The SD sensor measures the distance between the sensor and the underlying surface. These data contain continuous SD measurements including records on no-snow conditions with 1 h intervals. As the snow depth measurement at the hydro-climatic station site cannot represent the areal distribution of snow in the catchment, we also conducted a series of snow field measurements at 13 locations for two consecutive winters, 2016–2018, in order to understand the spatial variability of snow distribution and SD. In addition to the snow field measurements, we also observed surface temperatures for 2016–2017 using temperature iButtons at near surface (5 cm) depth with a 3 h interval. The iButtons were placed at all the snow field measurement locations with altitudes ranging from 1193 to 1612 m.a.s.l.

2.2. Methods

The main idea of the methodology applied in this study was to downscale the coarse resolution MODIS data based on a combination of other freely available optical remote sensing images, as well as ground measurements. An altitude-dependent SCD rate derived from MODIS data and DEM coupled with snow cover area (SCA) derived from the combination of Landsat and Sentinel images were used to produce annual SCD maps with 30 m spatial resolution from 2000–2017. We used the daily temporal resolution of MODIS Aqua and Terra satellites for deriving the temporal distribution of snow over the last 17 years, particularly the altitude dependent SCD rate, snow onset (S_o), and melt (S_m) dates. The Landsat and Sentinel images, with high spatial resolutions, were used for constructing the spatial distribution of snow in the study area. The abbreviations used in this study are shown in Table A1, Appendix A, and the flowchart of the procedure can be found in Figure 2.

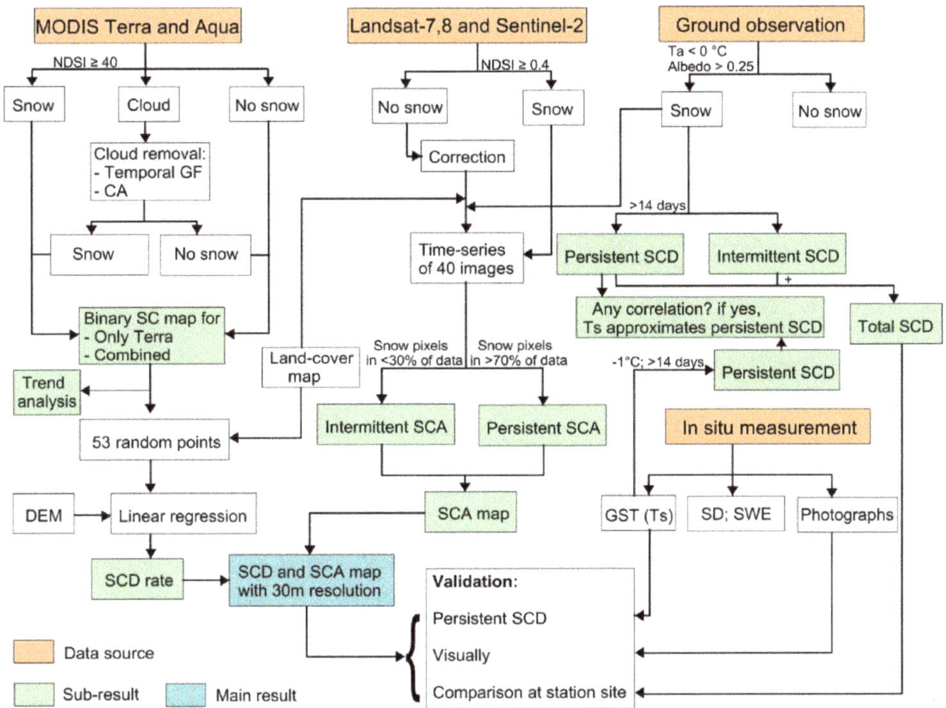

Figure 2. Flowchart of the procedure to map snow cover duration (SCD) and snow cover area (SCA) in the Sugnugur catchment. MODIS is Moderate Resolution Imaging Spectroradiometer. NDSI is normalized difference snow index. GF and CA are gap-filling and conditional adjustments, respectively. T_s and T_a denote daily mean ground surface temperature (GST) and daily mean air temperature, respectively. SD and SWE are snow depth and water equivalent, respectively.

For all satellite images, we adopted the normalized difference snow index (NDSI) with a threshold of ≥ 0.4 to denote snow cover [40,42–44].

$$\text{NDSI} = \frac{\text{Green} - \text{SWIR}}{\text{Green} + \text{SWIR}} \qquad (1)$$

where *Green* represents reflectance in a visible band and *SWIR* is reflectance in a short-wave infrared band.

Estimating SCD from ground temperature: The timing characteristic of snow cover was studied at the hydro-climatic station to understand the general snow accumulation and ablation processes in the study area. SCD is commonly derived from SD, SWE, or satellite data. In this study, a daily mean air temperature (T_a) of <0 °C and a daily mean albedo of ≥ 0.25 (unitless) were used to distinguish snow from the existing vegetation (Figure 3). We assume a persistent snow cover (continuous) if these two conditions are consistent for at least 14 days, otherwise it will be assigned as intermittent (not continuous). Occasional autumn and spring (transition periods) snowfalls develop intermittent snow cover lasting for a couple of days, whereas decreased air temperature towards the winter months (DJF) allows the snow cover to become persistent. We classified snow duration for each year as the sum of both intermittent SCD and persistent SCD.

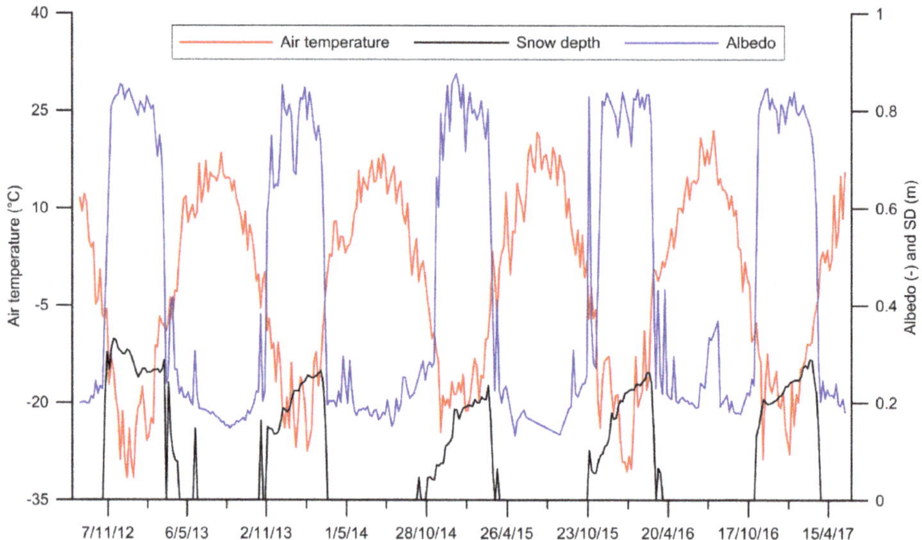

Figure 3. Snow depth (SD) collected from the SR50A-L sensor at the hydro-climatic station site by coupling with daily air temperature and albedo. The values are shown as 5-day averages.

We also used surface temperatures collected from the hydro-climatic station as proxies to classify persistent snow cover, such that if the daily mean surface temperature (T_s) remained below -1 °C for at least 14 days, the snow cover was determined as persistent. These conditions were then applied to our iButton measurements, which allowed us to validate our reconstructed SCD across the catchment. We examined the sign and strength of the relationship between the persistent SCD determined from the T_a and albedo and that approximated from T_s using Pearson's correlation. We assumed that if a positive and significant relationship existed, then persistent SCD may be reconstructed from T_s. Winter 2012/2013 was an exceptional year with relatively low temperatures in autumn, which led to a high number of below -1 °C days. Therefore, we did a manual correction by selecting the days between the date when first snowfall was observed after the surface temperature dropped below the threshold and the date when albedo dropped to below 0.25.

MODIS derived snow metrics: The timing characteristic of snow over the study area was analyzed using the daily MODIS Terra and Aqua data by determining snow onset (S_o) and snow melt (S_m) dates, as well as SCD for snow seasons 2000–2017. We first separated the cloudy pixels, designated with a value of 250, from the cloud free pixels and then detected the snow cover for the cloud free pixels using the NDSI to create daily 500 m snow maps.

To reduce cloud coverage, we applied a series of cloud removal approaches to Terra and the combination of Terra and Aqua data separately. First, we applied a 1-day backward and forward gap-filling (temporal GF) approach, with a limit of 3 days, to both Terra and Aqua such that when a pixel is cloudy, it goes one day backward and forward to test for a cloud free pixel [11,44]. Following this temporal gap-filling, a conditional adjustment (CA), which takes spatial and seasonality corrections into account, was applied. The CA assumes that if the given pixel after the temporal gap-filling is still cloudy, it can be reclassified as snow or no-snow under certain circumstances; if two of the 8-neighboring pixels at lower elevation are cloud free and the day of year is within the range between So and Sm dates of that certain pixel, we assigned snow. The So and Sm dates were determined for each MODIS pixel as the first and the last day of the classified snow cover days within the hydrologic year; it also defined the length of snow season for each pixel.

After adjusting for cloud cover, gap-filled images from both Terra and the combination data still underestimated SCD over heterogeneous topography and forested areas because the coverage area of a single 500×500 m MODIS image includes a high range of elevations as well as both forested north- and steep south-facing slopes. Thus, we generalized the annual number of SCD from the MODIS data for each year over the study area based on annual SCD rates. To do so, 53 random points were selected, either on mountain peaks or lowland areas where no forests or shrubs exist and the topography is homogenous, in order to employ linear regression analysis between the SCD from the improved MODIS data and DEM; later the regression parameters were used for deriving the SCD rate for each year.

Trend analysis using linear least squares regression were also conducted for snow onset (S_o), snowmelt (S_m) dates, and SCD from the improved MODIS daily snow-cover. Since air temperature is one of the main parameters that drives the snow accumulation and ablation processes, we calculated a trend analysis for monthly mean air temperature during the snow season, using observations from the nearest long-term climate station at Chinggis Khan International Airport (60 km south from the hydro-climatic station). The results can be found in Section 3.2.2.

Landsat and Sentinel images to correct for land-cover and topography: The generalized SCD is not realistic because it includes the intermittent SCA, which mostly occurs on the steep south-facing slopes and wind-blown surfaces. To identify these areas, we created a sequence of snow cover maps from the combination of Landsat-7, Landsat-8, and Sentinel-2A images (40 images in total) taken on clear-sky days. These time sequence maps are helpful in showing the development of SCA with high spatial resolution in the study area. Although the spatial resolutions of the satellites are high, snow cover below the forest canopy and dense shrub is difficult to detect [45]. Therefore, a correction method, which considers altitude dependency and land-cover types, was applied to all Landsat and Sentinel images to correct snow cover in those areas. More specifically, if a pixel from either a Landsat or Sentinel image was classified as snow at the hydro-climatic station, the areas in higher elevations with forests or shrubs were also classified as snow. If not, this assumption was ignored. By repeating this correction for each image, the change detection analysis was completed for the snow season period 2016–2017, representing the development of SCA in the catchment. Based on the changes in SCA, we separated the study area as persistent SCA (pixels classified as snow in >70% of the data) and intermittent SCA (pixels classified as snow in <30% of the data). The result can be found in Section 3.3. We assumed that the general spatial distribution of snow would be similar for each year to obtain the mean SCD for winters 2000/2001 to 2016/2017 over the seasonal persistent SCA by using the empirical altitude-dependent SCD rates that were found from the improved MODIS data.

Result assessment: To validate the results, we first compared the estimated total SCD with the observed total SCD at the hydro-climatic station. Daily surface temperature measurements (T_s) from the iButtons, distributed in the catchment, were also used by comparing the linear relationship, which was found at the hydro-climatic station, to that between the estimated persistent SCD days and days with $<-1\ ^\circ$C of T_s from the iButtons (Figure 2). Photographs from the time-lapse camera and snow field measurements were also used to verify the spatial distribution of snow visually.

3. Results

3.1. Seasonal Snow Cover from In Situ Measurements

In general, seasonal snow cover lasts from the beginning of October until the end of March with strong variability during the transition period. This transition period refers to late autumn and early spring with occasional intermittent snowfalls that develop a non-continuous snow cover that is relatively thin and temporary. Observed snow depth derived with the combination of $T_a < 0\ ^\circ$C and albedo ≥ 0.25 showed that the total SCD ranged from 145 to 157 days with a mean depth of 12–25 cm/year from 2012–2017. The intermittent and persistent snow cover days were distinguished for each year (Table 1). The duration of persistent snow cover was relatively stable while the intermittent snow cover days showed strong variability ranging from 6 to 21 days. The maximum snow depths were observed in the second half of March before the main melt started. There was no clear relationship between the variability of snow depth and the duration of snow cover.

Table 1. Intermittent and persistent snow cover days (SCD) during the observation period. [1] and [2] denote the persistent SCD derived from the combination of albedo and air temperature and below −1 °C of daily mean surface temperature, respectively.

	2012/13	2013/14	2014/15	2015/16	2016/17
Intermittent	13	12	13	21	6
Persistent	132 [1]/140 [2]	135 [1]/135 [2]	144 [1]/143 [2]	137 [1]/139 [2]	141 [1]/147 [2]
Total	145	146	157	156	147

As snow cover influences the surface energy balance significantly [46], the temperature at the underlying surface can serve as a proxy of absence and presence of seasonal persistent snow cover. Therefore, we validated persistent SCD using T_s at several locations, where snow cover was underestimated from MODIS sensors, and which is demonstrated in Figure 4 by plotting snow depth together with surface temperature measurements for each year at the hydro-climatic station.

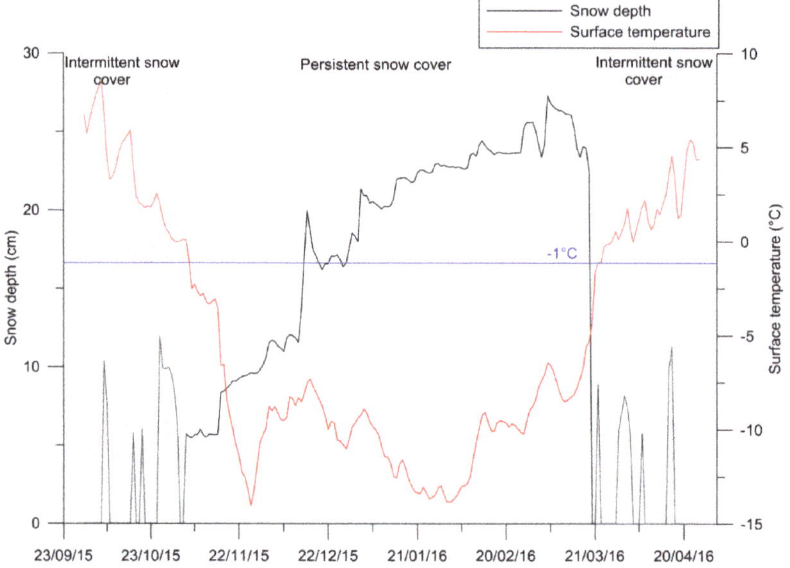

Figure 4. An example of the discrimination of intermittent and persistent snow cover, determined from the combination of albedo and air temperature measurements. The red and blue lines indicate the observed daily mean surface temperature and the threshold of −1 °C, respectively. The black line describes the typical snow accumulation during the snow season.

For the observation period at the hydro-climatic station, the distinguished persistent SCD showed a moderate and positive correlation (r = 0.64, p < 0.05) with the persistent SCD determined from the surface temperature record. These results suggest further that persistent SCD in other parts of the catchment could be verified using the surface temperature iButton measurements. Snow field measurements were conducted at each iButton location, and south-facing slopes were snow free while both valley bottom and north-facing slopes had similar SD, indicating no changes with increasing altitude (see details of the snow field measurements in Table S2, Supplementary Materials).

3.2. Temporal Distribution of Snow

3.2.1. Cloud Reduction

To reduce cloud cover in the daily Terra and the combined Aqua and Terra images, the temporal gap-filling (GF) and conditional adjustment (CA) were applied to both datasets. Figure 5 shows the results as the mean cloud coverage. The most obvious decrease appeared after the temporal GF filtering [47]. The relative percentage of cloud cover for the entire study area decreased up to 100% from the initial MODIS data to the final CA filtering. The combination of the two initial data yielded only slight improvements in reducing cloud coverage because the time difference between the two overpasses was about 3 h; hence general atmospheric conditions remained similar. The difference between Terra and the combined version after the final CA was not significant.

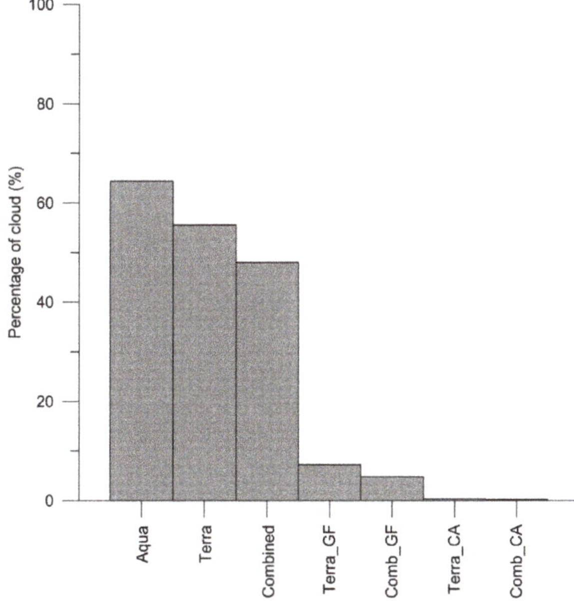

Figure 5. Changes in percent of cloud coverage after a series of cloud reduction steps applied to daily MODIS snow products Aqua and Terra; Combined—The combination of both satellites; Terra_GF—Gap-filling applied to only Terra; Comb_GF—Gap-filling applied to the combination; Terra_CA—Conditional adjustment applied to Terra_GF; Comb_CA—Conditional adjustment applied to Comb_GF.

3.2.2. SCD and Trend Analysis

Altitude, latitude, and solar radiation are known factors that govern SCD as well as their spatial distribution [48,49]. Due to the study area's relatively small size, stretched from east to west, the derived SCD from the improved MODIS data showed an insignificant relationship with potential solar radiation ($p = 0.33$); therefore, we ignored the influence of latitude and topography, and considered only the altitude dependency for deriving the SCD rate (days/m). The mean annual SCD ranged from 124–226 days from the lowland plain area to the mountain peak, showing clear underestimation over the forested and heterogeneous topography area (Figure 6). The mean increasing SCD rate was +6 days/100 m with mean coefficient of determination $r^2 = 0.92$ (ranging from 0.71–0.96) with the probability value $p < 0.001$, and it agrees well with the result (+5.9 days/100 m) found in other Central Asian mountains [7].

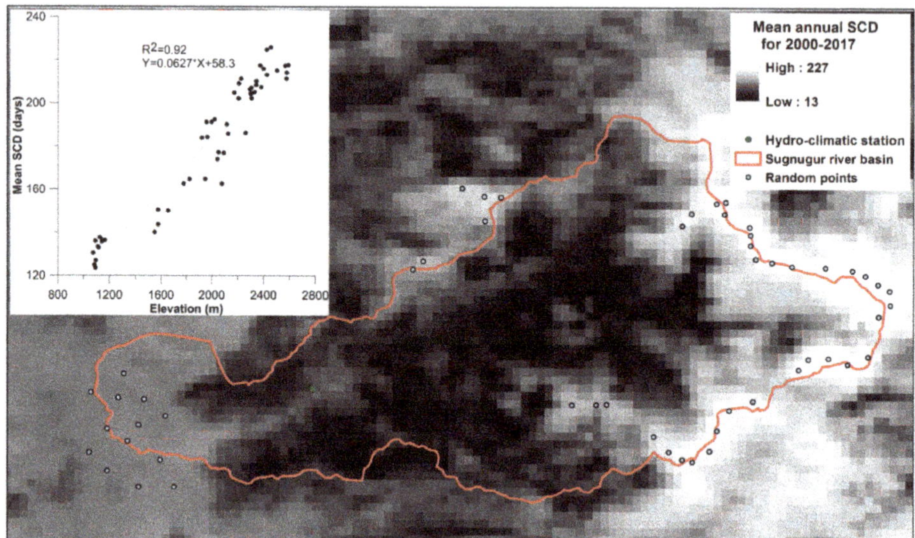

Figure 6. Mean annual snow cover duration (SCD) for 2000–2017 from the improved MODIS-Terra and the derived mean SCD rate (inset image). Green and blue dots denote the locations of hydro-climatic station and selected random points, respectively. Areas with heterogeneous topography and forest cover mostly in the middle part of the catchment demonstrate significant underestimation, such as only 13 days, as shown in the scale.

The trend analyses of MODIS derived snow metrics (S_o, S_m, and SCD) and long-term air temperature of snow season (from October to April) are shown in Figure 7. The S_o and S_m dates exhibited large inter-annual variability with insignificant slightly increasing delays in both S_o (p = 0.53) and S_m (p = 0.77) dates, indicating a slight shift in the snow season. To implement trend analysis for the total SCD, we separated the study area into two parts (mountain and plain area) based on an elevation threshold (2300 m.a.s.l) of the tree line [25], excluding the area that is always underestimated from the MODIS data because of the forest canopies and topographic heterogeneity. Overall, the SCD over both the mountains and the plains showed insignificant decreasing trends (p = 0.96 for plains and p = 0.41 for mountains). The mean change rates were −0.5 days/year for mountains and −0.04 days/year for the plains, respectively. The mean SCD for mountain peaks dropped down to its minimum in winter 2014/2015 and increased to its maximum in winter 2012/2013. On the contrary, SCD in the plains showed high variability between 2000 and 2008, and then it stabilized. The hydrologic year 2001/2002 had the shortest SCD in the plains due to late snow cover onset and early snow melt, and is in good agreement with the other findings in Central Asia [50]. In addition, there was no decline in snow cover over our study period, which is also in an agreement with other studies in the region, e.g., [1]. Nevertheless, the time series of 17 years may be relatively short to produce a trend analysis of these snow metrics, and a longer time-series of data may be necessary for a more comprehensive analysis.

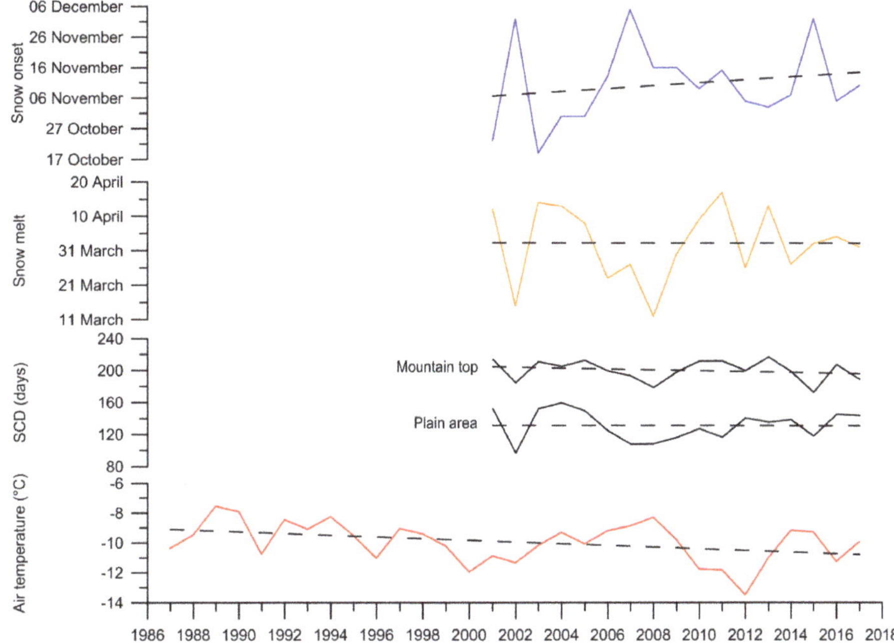

Figure 7. Trend analysis for the mean snow onset, snow melt, and SCD for 2001–2017, as well as monthly mean air temperature during snow season (ONDJFMA). Mountain peaks and plain areas were separated for SCD analysis. The temperature trend was analyzed using daily mean values from the nearest long-term observational site at the Chinggis Khan International Airport. There was no significant trend in each snow metric, while a slightly decreasing trend ($p < 0.05$) in air temperature was detected during the snow season, including intermittent and persistent snow periods.

Interestingly, the monthly mean air temperature during the snow seasons, including intermittent and persistent snow periods, showed a decreasing trend over the last three decades. This decrease was statistically significant ($p < 0.05$), but only for the mid-season months (DJF) (Table 2). Unlike in many regions in the world, a decrease or no increase in winter air temperature has been observed since the 1990s, replacing the general increasing trend since the 1960s over Central Asia under the influence of weakening SH [9,10,49]. The mean temperature of the snow season also showed an insignificant negative relationship ($r^2 = 0.19$, $p = 0.07$) with the SCD on the mountain top, indicating the decreasing temperature may have played an important role in preventing SCD decrease.

Table 2. The significance of trend analysis in monthly mean air temperature over the last 31 years at the Chinggis Khan International Airport (60 km from the study area).

	r^2	Slope	Intercept	p Value
October	0.002	−0.0092	−0.2744	0.78
November	0.053	−0.0721	−11.26	0.25
December	0.22	−0.142	−18.06	0.012 *
January	0.19	−0.14	−20.8	0.023 *
February	0.18	−0.15	−15.5	0.021 *
March	0.0009	−0.009	−7.82	0.09
April	0.03	0.0435	1.42	0.32

* Significant, when $p < 0.05$

3.3. Spatial Distribution of Snow

For analyzing the spatial distribution of snow in the study area, we exploited the high spatial resolutions of Landsat-7, Landsat-8, and Sentinel-2A images to create a sequence of snow cover maps during the snow season 2016/2017 (Figure S1, Supplementary Materials). The first snow covered area was captured on 11 October 2016 over mountain peaks by Landsat-7, and then decreased by 16 October after 5 days as captured by Sentinel-2 during the autumn transition period. The percent of SCA in the study area increased following several snow events and eventually reached 99.6% by 21 November, including the hydro-climatic station site (Figure 8). The persistency of seasonal snow cover for the Sugnugur catchment began after the heavy snow events from 6–7 November 2016.

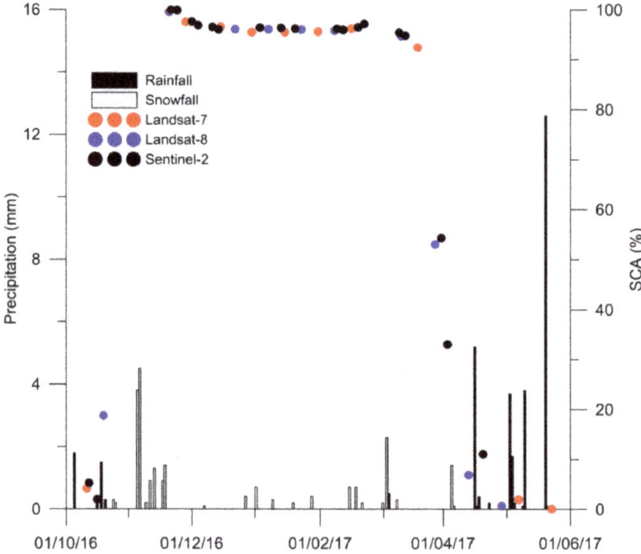

Figure 8. Precipitation events, distinguished as rain or snow, observed at the hydro-climatic station using the combination of albedo and air temperature measurements, and the inter-annual variability of SCA detected by different satellites over the Sugnugur catchment for winter 2016/2017.

We were able to use 28 images (70% of the total data) to identify the spatial distribution of seasonal persistent snow cover in 94.2% of the total area. The intermittent and snow free areas were 5.3% and 0.5% of the catchment area during the winter 2016/2017, respectively. The different spatial resolution of the two sensors may affect their combined use of time-series analysis [51] and result in minor uncertainties, mostly in the intermittent SCA. The intermittent snow areas appeared mostly on the steep south-facing slopes. Following several snow events, the south-facing slopes were covered by snow, but shortly became snow free due to higher sublimation rates or blowing-wind [52]. However, the intermittent SCD on the south-facing slopes still remain unclear because of the coarse spatial resolution of MODIS and non-daily coverage of the combined Landsat and Sentinel overpasses. The absence of snow cover on south-facing slopes was captured by our time-lapse camera, and was also visually inspected during the snow field campaigns (Figure S2 and Table S2, Supplementary Materials).

3.4. Mapping Snow Cover

Finally, we created a snow duration map (Figure 9) with 30 m resolution based on the findings that included the combination of both ground measurements and various optical remote sensing images. The map was produced only for seasonal persistent SCA because SCD for intermittent SCA had higher uncertainty. The simple correction, which was applied to Landsat and Sentinel images using the supervised land-cover classification map, improved the spatial accuracy of SCA that was not detectable or often underestimated by MODIS satellites.

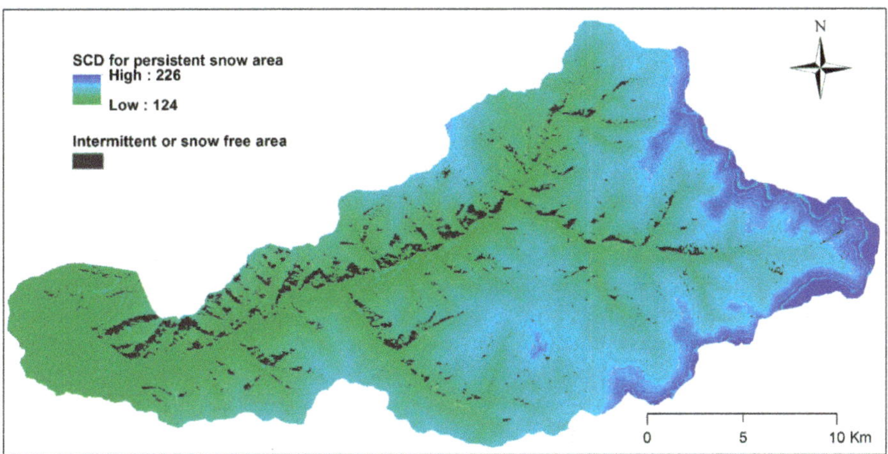

Figure 9. The total snow cover duration (SCD) days in Sugnugur catchment calculated only for persistent snow cover area. The intermittent and snow free areas were excluded due to their unknown snow accumulation periods.

The mean altitude-dependent SCD rate +6 days/100 m was consistent with that found in the same region [7]. The south-facing slopes were mostly snow free or with intermittent snow cover during winter due to high sublimation rate, wind blowing, and slope steepness. The mean annual SCD ranged from 124–226 days from the lowland plain area to the mountain peak (elevation difference of ~1700 m and horizontal distance of ~35 km), or showing high variability for such a small scale. Therefore, snow-melt water from the higher elevations with extended snow season is extremely important for balancing the Sugnugur River discharge during spring and autumn dry seasons.

To match our temporal record of snow depth observations at the hydro-climatic station, we compared the estimated SCD with the total observed SCD for the winters of 2012/2013 to 2016/2017. There was a clear underestimation of satellite derived SCD and resulted in an error of 12–13 days/year (Figure S3, Supplementary Materials). Previous studies have described MODIS snow products in clear-sky conditions to return high accuracies of >93% for Terra and >90% for Aqua [28,53]. Assuming these accuracies, our underestimation equates to 9% and is within the MODIS accuracy range. Nevertheless, errors are likely propagated through the relatively weak snow detection rate of MODIS sensors during the transition periods [13,54], the position of the hydro-climatic station relative to the receipt of incoming solar radiation, as well as the accuracy of annual altitude-dependent SCD rate.

The absence of snow cover on the south-facing slopes was captured by the automated time-lapse camera system, as well as was visually observed during fieldwork, which took place in the beginning of March 2017 and 2018, or the accumulated persistent snow cover period. The photographs from the camera showed a light snow event on 5 March 2017 together with the previous and following days, indicating the south-facing slopes were snow-free before the event (Figure S2, Supplementary Materials). The snow field measurements showed similar snow depths on both valley bottom and north-facing slope without any altitude-dependent increase. Therefore, we speculate that snow depth was homogeneous for the seasonal persistent SCA over the study area.

The days with below −1 °C of T_s at the hydro-climatic station showed a positive relationship with the observed persistent SCD days. This suggests the possibility of validating the estimated SCD against the days with $T_s < -1$ °C in the upper part of the catchment. For this analysis, surface temperature measurements at the valley bottom and north-facing slopes in different land-cover types were used because the south-facing slopes were snow-free, and thus were excluded from the final SCD map. The estimated SCD determined from T_s revealed good agreement with an r^2 of 0.85 (p < 0.001) and a mean bias of ± 1 day compared to the hydro climatic station site. This would mean that the mean altitude-dependent SCD rate is consistent for persistent snow cover areas where MODIS snow data are contaminated by the topographic heterogeneity, forest, and shrubs.

4. Discussion

Detection of snow from MODIS data has proven to exceed 90% accuracy on clear-sky days [44,48,53], but cloud coverage remains a major problem for all optical satellites. The temporal GF and CA (similar to those in [11,44]) applied to the MODIS data resulted in promising cloud reductions of up to 100%. This success is largely attributed to the SH, which enables a high number of clear-sky days over Mongolia (Figure 5). The obtained S_o, S_m, and SCD did not indicate any significant trends since 2000, while mid-season winter (DJF) air temperature showed a decreasing trend over the last three decades (Figure 7) associated with the weakening of SH [8–10]. The SH starts to form in early autumn and causes early forming snow cover due to its contribution to changes in temperature and precipitation [6,8,9]. Therefore, the increased delay (statistically not significant) in S_o might be attributed to the seasonal strengthening of the SH. However, more detailed investigations on extreme weather and precipitation events in conjunction with the SH are necessary to draw strong conclusions.

In this study, we applied a simple alternative altitude-dependent correction to Landsat and Sentinel images to produce a time series of snow distribution in the Sugnugur catchment. Results showed the development of snow cover during the snow season in time and space and the classified seasonal intermittent and persistent SCAs (Figure 9). Classifying persistent SCA from intermittent SCA is important for more accurate hydrologic and water balance modelling. Since the temporal resolution of the produced snow maps from Landsat and Sentinel images is still coarse, SCD for intermittent SCA remains uncertain, and thus further investigations should also include the timing of intermittent snow cover where ground measurements are not available.

We also demonstrated the altitude-dependent SCD rate to be +6 days/100 m and is in good agreement with other rates across other Central Asian Mountains [7]. The overall accuracy of SCD was ~91% with the mean underestimation of ~12–13 days/year, which was probably caused by the reduced MODIS snow detection rate for thin occasional intermittent snow cover that occurs in autumn and spring months. The snow field measurements supported more comprehensive snow metrics, such as SWE and SD in the study area. In general, no changes in SD with increasing elevation were evident over the catchment. However, SD can vary at local scales because of forest interception, snow redistribution, as well as different sublimation rates [55], and it should be studied in future investigations.

As snow cover contributes to the surface energy balance [56], it appears that the daily mean surface temperature (T_s) of $< -1\ ^\circ$C could be a good approximation of the overlaying existing persistent snow cover. Nevertheless, the applicability of using T_s for defining seasonal persistent SCD might be limited to certain spatial extents in cold regions because of the high climatic variability and characteristics of snow cover in the high altitudes and latitudes of the Northern Hemisphere. The T_s from iButtons also indicated that the changes in land-cover alter the timing of snow-melt processes, showing an earlier S_m in burned forest relative to unburned forest [57]. Therefore, the consequent effect of the earlier S_m in conjunction with energy exchange and long term snow monitoring, together with continuous river discharge measurement, should be discussed in further studies for observing the contribution of snow-melt water in the regional water resources.

5. Conclusions

The detailed spatio-temporal distribution of snow has not been previously studied in Mongolia. In this study we examined snow cover in the Khentii Mountains, which is an important headwater for Lake Baikal and is the origin of the Selenga River. The use of multi-source data provided us essential information for understanding the timing and spatial characteristics of snow in the Sugnugur catchment within a semi-arid climate where snow plays an important role in the regional water resources and surface energy balance.

The snow ground observation at the hydro-climatic station demonstrated strong variability in the timing of snow, especially during the intermittent snow cover period, showing a duration of 6–21 days/year, while persistent SCD was relatively stable with 132–141 days. Daily mean surface temperature of below $-1\ ^{\circ}$C with continuation of at least 14 days could give a good proxy of seasonal persistent snow cover. The manually measured SD at different elevations and land-cover types did not indicate any variability during the two consecutive snow field measurements.

The combined MODIS, Landsat, and Sentinel retrievals with various spatial and temporal resolutions gave us an opportunity to map snow cover and its duration in the mountainous region with 30 m resolution by applying a series of adjustments, including temporal gap-filling and conditional adjustments. The derived SCD rate is similar to that found in other Central Asian Mountains. The derived annual SCD shows high variations, but no significant trend since 2000. Overall, the combination of snow ground observation, field measurements, and high resolution open source optical remote sensing images with high temporal resolutions can be an option for understanding snow distribution and duration in mid-latitude mountain regions.

Supplementary Materials: The following are available online at http://www.mdpi.com/2076-3263/9/1/53/s1, Table S1. The acquisition dates of the satellite images that were used for producing the spatial distribution of snow. All selected images were taken on cloud-free conditions. Table S2. The overview of snow field measurements which were conducted on March 1, 2017 and 2018. Figure S1. Development of SCA for winter 2016-2017 from the combination of Landsat and Sentinel retrievals. The light-blue color indicates snow cover. Figure S2. Typical snow distribution in the Sugnugur catchment during the end of seasonal persistent snow cover period. Photographs were taken using a normal digital camera with automatic time-lapse setting. Figure S3. Anomalies of the estimated SCD for Terra and the combined version compared to the observed total SCD at the hydro-climatic station site for winters 2012/2013 to 2016/2017.

Author Contributions: M.M. analyzed the in situ and Landsat data, proposed the methodology, and wrote the manuscript. S.G. analyzed the MODIS and Sentinel data and contributed to the methodology and review. C.G.P. provided crucial suggestions and improved the manuscript writing. G.Y. and J.Y. carried out the snow field measurements. L.M. provided climatic data, supported the fieldwork, and supervised the study.

Funding: This research received no external funding.

Acknowledgments: Munkhdavaa Munkhjargal received funding from the Federal Ministry of Education and Research (BMBF) through the German Academic Exchange Service (DAAD) during the preparation of this manuscript. We acknowledge financial support by Deutsche Forschungsgemeinschaft within the funding programme Open Access Publishing, by the Baden-Württemberg Ministry of Science, Research and the Arts and by Ruprecht-Karls-Universität Heidelberg.

Conflicts of Interest: The authors declare no conflict of interest.

Appendix A

Table A1. Abbreviations used in the paper and their explanation.

Abbreviation	Explanation	Unit
SCD	Snow cover duration	days
SCD RATE	Snow cover duration rate	day/m
SCA	Snow cover area	%
SWE	Snow water equivalent	mm
S_O	Snow onset	date
S_M	Snow melt	date
SH	Siberian high	-
DEM	Digital elevation model	m
GF	Gap-filling	-
CA	Conditional adjustment	-
NDSI	Normalized difference snow index	-
T_A	Daily average air temperature	$^{\circ}$C
T_S	Daily average ground surface temperature	$^{\circ}$C

References

1. Wang, X.; Wu, C.; Wang, H.; Gonsamo, A.; Liu, Z. No evidence of widespread decline of snow cover on the Tibetan Plateau over 2000–2015. *Sci. Rep.* **2017**, *7*, 14645. [CrossRef] [PubMed]
2. Viviroli, D.; Dürr, H.H.; Messerli, B.; Meybeck, M.; Weingartner, R. Mountains of the world, water towers for humanity: Typology, mapping, and global significance. *Water Resour. Res.* **2007**, *43*, 1–13. [CrossRef]
3. Chevallier, P.; Pouyaud, B.; Mojaïsky, M.; Bolgov, M.; Olsson, O.; Bauer, M.; Froebrich, J. River flow regime and snow cover of the Pamir Alay (Central Asia) in a changing climate. *Hydrol. Sci. J.* **2014**, *59*, 1491–1506. [CrossRef]
4. Kim, Y.; Kimball, J.S.; Robinson, D.A.; Derksen, C. New satellite climate data records indicate strong coupling between recent frozen season changes and snow cover over high northern latitudes. *Environ. Res. Lett.* **2015**, *10*, 084004. [CrossRef]
5. Callaghan, T.V.; Johansson, M.; Brown, R.D.; Groisman, P.Y.; Labba, N.; Radionov, V.; Barry, R.G.; Bulygina, O.N.; Essery, R.L.H.; Frolov, D.M.; et al. The Changing Face of Arctic Snow Cover: A Synthesis of Observed and Projected Changes. *AMBIO* **2011**, *40*, 17–31. [CrossRef]
6. Bednorz, E.; Wibig, J. Circulation patterns governing October snowfalls in southern Siberia. *Theor. Appl. Climatol.* **2017**, *128*, 12–139. [CrossRef]
7. Zhou, H.; Aizen, E.; Aizen, V. Seasonal snow cover regime and historical change in Central Asia from 1986 to 2008. *Glob. Planet. Chang.* **2017**, *148*, 192–216. [CrossRef]
8. Cohen, J.; Foster, J.; Barlow, M.; Saito, K.; Jones, J. Winter 2009–2010: A case study of an extreme Arctic Oscillation event. *Geophys. Res. Lett.* **2010**, *37*. [CrossRef]
9. Cohen, J.; Screen, J.A.; Furtado, J.C.; Barlow, M.; Whittleston, D.; Coumou, D.; Francis, J.; Dethloff, K.; Entekhabi, D.; Overland, J.; et al. Recent Arctic amplification and extreme mid-latitude weather. *Nat. Geosci.* **2014**, *7*, 627–637. [CrossRef]
10. Panagiotopoulos, F.; Shahgedanova, M. Observed Trends and Teleconnections of the Siberian High: A Recently Declining Center of Action. *J. Clim.* **2005**, *18*, 1411–1422. [CrossRef]
11. Lindsay, C.; Zhu, J.; Miller, A.E.; Kirchner, P.; Wilson, T.L. Deriving Snow Cover Metrics for Alaska from MODIS. *Remote Sens.* **2015**, *7*, 12961–12985. [CrossRef]
12. Notarnicola, C.; Duguay, M.; Moelg, N.; Schellenberger, T.; Tetzlaff, A.; Monsorno, R.; Costa, A.; Steurer, C.; Zebisch, M. Snow Cover Maps from MODIS Images at 250 m Resolution. *Remote Sens.* **2013**, *5*, 110–126. [CrossRef]
13. Rittger, K.; Painter, T.H.; Dozier, J. Advances in Water Resources Assessment of methods for mapping snow cover from MODIS. *Adv. Water Resour.* **2013**, *51*, 367–380. [CrossRef]
14. Walters, R.D.; Watson, K.A.; Marshall, H.; Mcnamara, J.P.; Flores, A.N. A physiographic approach to downscaling fractional snow cover data in mountainous regions. *Remote Sens. Environ.* **2014**, *152*, 413–425. [CrossRef]
15. Gafurov, A.; Vorogushyn, S.; Farinotti, D.; Duethmann, D.; Merkushkin, A.; Merz, B. Snow-cover reconstruction methodology for mountainous regions based on historic in situ observations and recent remote sensing data. *Cryosphere* **2015**, *9*, 451–463. [CrossRef]
16. Gurung, D.R.; Maharjan, S.B.; Shrestha, A.B.; Shrestha, M.S.; Bajracharya, S.R.; Murthy, M.S.R. Climate and topographic controls on snow cover dynamics in the Hindu Kush Himalaya. *Int. J. Climatol.* **2017**, *37*, 3873–3882. [CrossRef]
17. Liu, J.P.; Zhang, W.C. Long term spatio-temporal analyses of snow cover in Central Asia using ERA-Interim and MODIS products. *IOP Conf. Ser. Earth Environ. Sci.* **2017**, *57*, 012033. [CrossRef]
18. Manuel, G. An Operational Snow Cover Product from Sentinel-2 and Landsat-8 Data for Mountain Regions. *La Montagne, Territoire D'Innovation*; Grenoble, France, January 2017. Available online: https://hal.archives-ouvertes.fr/halshs-01486820 (accessed on 9 January 2018).
19. Dedieu, J.; Carlson, B.Z.; Bigot, S.; Sirguey, P.; Vionnet, V.; Choler, P. On the Importance of High-Resolution Time Series of Optical Imagery for Quantifying the Effects of Snow Cover Duration on Alpine Plant Habitat. *Remote Sens.* **2016**, *8*, 481. [CrossRef]
20. Selkowitz, D.; Rittger, K. Developing a 30 m daily snow covered area time series for the Sierra Nevada Alpine using Landsat and MODIS data. In Proceedings of the AGU Fall Meeting, San Francisco, CA, USA, 3–7 December 2012.

21. Li, H.Y.; He, Y.Q.; Hao, X.H.; Che, T.; Wang, J.; Huang, X.D. Downscaling Snow Cover Fraction Data in Mountainous Regions Based on Simulated Inhomogeneous Snow Ablation. *Remote Sens.* **2015**, *7*, 8995–9019. [CrossRef]

22. Cristea, N.C.; Breckheimer, I.; Raleigh, M.S.; HilleRisLambers, J.; Lundquist, J.D. An evaluation of terrain-based downscaling of fractional snow covered area data sets based on LiDAR-derived snow data and orthoimagery. *Water Resour. Res.* **2017**, *53*, 6802–6820. [CrossRef]

23. Kostadinov, T.S.; Lookingbill, T.R. Remote Sensing of Environment Snow cover variability in a forest ecotone of the Oregon Cascades via MODIS Terra products. *Remote Sens. Environ.* **2015**, *164*, 155–169. [CrossRef]

24. Kopp, B.J.; Minderlein, S.; Menzel, L. Soil Moisture Dynamics in a Mountainous Headwater Area in the Discontinuous Permafrost Zone of northern Mongolia. *Arct. Antarct. Alp. Res.* **2014**, *46*, 459–470. [CrossRef]

25. Kopp, B.J.; Lange, J.; Menzel, L. Effects of wildfire on runoff generating processes in northern Mongolia. *Reg. Environ. Chang.* **2016**, *17*, 1951–1963. [CrossRef]

26. Zhang, Y.; Mamoru, I.; Tetsou, O.; Dambaravja, O. Role of snow playing in water cycle in semi- arid region of Mongolia. In Proceedings of the 3rd International Workshop on Terrestrial Change in Mongolia, Tsukuba, Japan, 9–10 November 2004; pp. 23–24.

27. Wimmer, F.; Schlaffer, S.; aus der Beek, T.; Menzel, L. Distributed modelling of climate change impacts on snow sublimation in Northern Mongolia. *Adv. Geosci.* **2009**, *21*, 117–124. [CrossRef]

28. Parajka, J.; Holko, L.; Kostka, Z.; Blöschl, G. MODIS snow cover mapping accuracy in a small mountain catchment-comparison between open and forest sites. *Hydrol. Earth Syst. Sci.* **2012**, *16*, 2365–2377. [CrossRef]

29. Klein, A.G.; Barnett, A.C. Validation of daily MODIS snow cover maps of the Upper Rio Grande River Basin for the 2000–2001 snow year. *Remote Sens. Environ.* **2003**, *86*, 162–176. [CrossRef]

30. Tekeli, A.E.; Akyu, Z.; Arda, A.S.; Sensoy, A.; Sorman, Ü. Using MODIS snow cover maps in modeling snowmelt runoff process in the eastern part of Turkey. *Remote Sens. Environ.* **2005**, *97*, 216–230. [CrossRef]

31. Parajka, J.; Blöschl, G. Validation of MODIS snow cover images over Austria. *Hydrol. Earth Syst. Sci.* **2006**, *10*, 679–689. [CrossRef]

32. Ault, T.W.; Czajkowski, K.P.; Benko, T.; Coss, J.; Struble, J.; Spongberg, A.; Templin, M.; Gross, C. Validation of the MODIS snow product and cloud mask using student and NWS cooperative station observations in the Lower Great Lakes Region. *Remote Sens. Environ.* **2006**, *105*, 341–353. [CrossRef]

33. Wang, X.; Xie, H.; Liang, T. Evaluation of MODIS snow cover and cloud mask and its application in Northern Xinjiang, China. *Remote Sens. Environ.* **2008**, *112*, 1497–1513. [CrossRef]

34. Liang, T.; Zhang, X.; Xie, H.; Wu, C.; Feng, Q.; Huang, X.; Chen, Q. Remote Sensing of Environment Toward improved daily snow cover mapping with advanced combination of MODIS and AMSR-E measurements. *Remote Sens. Environ.* **2008**, *112*, 3750–3761. [CrossRef]

35. Huang, X.; Liang, T.; Zhang, X.; Guo, Z. Validation of MODIS snow cover products using Landsat and ground measurements during the 2001–2005 snow seasons over northern Xinjiang, China. *Int. J. Remote Sens.* **2011**, *32*, 133–152. [CrossRef]

36. Gafurov, A.; Kriegel, D.; Vorogushyn, S.; Merz, B. Evaluation of remotely sensed snow cover product in Central Asia. *Hydrol. Res.* **2013**, *44*, 506–522. [CrossRef]

37. Dong, C.; Menzel, L. Improving the accuracy of MODIS 8-day snow products with in situ temperature and precipitation data. *J. Hydrol.* **2016**, *534*, 466–477. [CrossRef]

38. Dong, C.; Menzel, L. Producing cloud-free MODIS snow cover products with conditional probability interpolation and meteorological data. *Remote Sens. Environ.* **2016**, *186*, 439–451. [CrossRef]

39. Dong, C. Remote sensing, hydrological modeling and in situ observations in snow cover research: A review. *J. Hydrol.* **2018**, *561*, 573–583. [CrossRef]

40. Hall, D.K.; Riggs, G.A. *MODIS/Terra Snow Cover Daily L3 Global 500 m Grid*, version 6; NASA National Snow and Ice Data Center Distributed Active Archive Center: Boulder, CO, USA, 2016.

41. National Aeronautic and Space Administration (NASA) and Ministry of Economy, Trade, and Industry (METI). ASTER GDEM 2; U.S Geological Survey: Sioux Falls, South Dakota, 2011. Available online: https://earthexplorer.usgs.gov/ (accessed on 25 December 2017).

42. Landsat-7, Landsat-8 and Sentinel-2 (ESA) Image Courtesy of the U.S. Geological Survey. Available online: https://earthexplorer.usgs.gov/ (accessed on 15 November 2017).

43. Hall, D.K.; Riggs, G.A.; Salomonson, V.V.; Digirolamo, N.E.; Bayr, K.J. MODIS snow-cover products. *Remote Sens. Environ.* **2002**, *83*, 181–194. [CrossRef]

44. Hall, D.K.; Riggs, G.A.; Foster, J.L.; Kumar, S.V. Development and evaluation of a cloud-gap-filled MODIS daily snow-cover product Part of the Physical Sciences and Mathematics Commons. *Remote Sens. Environ.* **2010**, *114*, 496–503. [CrossRef]

45. Wang, X.; Wang, J.; Jiang, Z.; Li, H.; Hao, X. An Effective Method for Snow-Cover Mapping of Dense Coniferous Forests in the Upper Heihe River Basin Using Landsat Operational Land Imager Data. *Remote Sens.* **2015**, *7*, 17246–17257. [CrossRef]

46. Ishikawa, M.; Iijima, Y.; Zhang, Y.; Kadota, T.; Yabuki, H.; Ohata, T.; Battogtokh, D.; Sharkhuu, N. Comparable energy balance measurements on the permafrost and immediate adjacent permafrost-free slopes at the southern boundary of Eurasian permafrost, Mongolia. In Proceedings of the Ninth International Conference on Permafrost, Fairbanks, AK, USA, 29 June–3 July 2008.

47. Hüsler, F.; Jonas, T.; Riffler, M.; Musial, J.P.; Wunderle, S. A satellite-based snow cover climatology (1985–2011) for the European Alps derived from AVHRR data. *Cryosphere* **2014**, *8*, 73–90. [CrossRef]

48. Liu, J.; Zhang, W.; Liu, T. Monitoring recent changes in snow cover in Central Asia using improved MODIS snow-cover products. *J. Arid Land* **2017**, *9*, 763–777. [CrossRef]

49. Hu, Z.; Zhang, C.; Hu, Q.; Tian, H. Temperature Changes in Central Asia from 1979 to 2011 Based on Multiple Datasets. *J. Clim.* **2014**, 1143–1167. [CrossRef]

50. Dietz, A.J.; Keunzer, C.; Conrad, C. Snow-cover variability in Central Asia between 2000 and 2001 derived from improved MODIS daily snow-cover products. *Int. J. Remote Sens.* **2013**, *34*, 3879–3902. [CrossRef]

51. Mandanici, E.; Bitelli, G. Preliminary Comparison of Sentinel-2 and Landsat 8 Imagery for a Combined Use. *Remote Sens.* **2016**, *8*, 1014. [CrossRef]

52. Revuelto, J.; Gilaberte, M. The effect of slope aspect on the response of snowpack to climate warming in the Pyrenees. *Theor. Appl. Climatol.* **2014**, *117*, 207–219.

53. Hall, D.K.; Riggs, G.A. Accuracy assessment of the MODIS snow products. *Hydrol. Process.* **2007**, *21*, 1534–1547. [CrossRef]

54. Pu, Z.; Xu, L.; Salomonson, V.V. MODIS/Terra observed seasonal variations of snow cover over the Tibetan Plateau. *Geophys. Res. Lett.* **2007**, *34*, L067606. [CrossRef]

55. Woo, M.; Marsh, P. Snow, frozen soils and permafrost hydrology in Canada, 1999–2002. *Hydrol. Process.* **2005**, *19*, 215–229. [CrossRef]

56. Zhang, T. Influence of the seasonal snow cover on the ground thermal regime: An overview. *Rev. Geophys.* **2005**, *43*, RG4002. [CrossRef]

57. Burles, K.; Boon, S. Snowmelt energy balance in a burned forest plot, Crowsnest Pass, Alberta, Canada. *Hydrol. Process.* **2011**, *25*, 3012–3029. [CrossRef]

Article

Discriminating Wet Snow and Firn for Alpine Glaciers Using Sentinel-1 Data: A Case Study at Rofental, Austria

Achim Heilig [1,2,*], Anna Wendleder [3], Andreas Schmitt [3,4] and Christoph Mayer [2]

[1] Department of Earth and Environmental Sciences, Munich University, 80539 Munich, Germany
[2] Geodesy and Glaciology, Bavarian Academy of Sciences and Humanities, 80539 Munich, Germany; Christoph.Mayer@keg.badw.de
[3] German Aerospace Center (DLR), German Remote Sensing Data Center (DFD), Land Surface Applications, 82234 Oberpfaffenhofen, Germany; Anna.Wendleder@dlr.de
[4] Department of Geoinformatics, University of Applied Sciences, 80335 Munich, Germany; andreas.schmitt@hm.edu
* Correspondence: heilig@r-hm.de; Tel.: +49-89-23031-1195

Received: 10 December 2018; Accepted: 22 January 2019; Published: 30 January 2019

Abstract: Continuous monitoring of glacier changes supports our understanding of climate related glacier behavior. Remote sensing data offer the unique opportunity to observe individual glaciers as well as entire mountain ranges. In this study, we used synthetic aperture radar (SAR) data to monitor the recession of wet snow area extent per season for three different glacier areas of the Rofental, Austria. For four glaciological years (GYs, 2014/2015–2017/2018), Sentinel-1 (S1) SAR data were acquired and processed. For all four GYs, the seasonal snow retreated above the elevation range of perennial firn. The described processing routine is capable of discriminating wet snow from firn areas for all GYs with sufficient accuracy. For a short in situ transect of the snow—firn boundary, SAR derived wet snow extent agreed within an accuracy of three to four pixels or 30–40 m. For entire glaciers, we used optical remote sensing imagery and field data to assess reliability of derived wet snow covered area extent. Differences in determination of snow covered area between optical data and SAR analysis did not exceed 10% on average. Offsets of SAR data to results of annual field assessments are below 10% as well. The introduced workflow for S1 data will contribute to monitoring accumulation area extent for remote and hazardous glacier areas and thus improve the data basis for such locations.

Keywords: SAR; transient snowline; annual AAR; mass balance; Rofental glaciers

---

1. Introduction

Changes in glacier mass balance are commonly used as indicators of global climate change [1]. However, contrary to central Europe or Scandinavia, regular glacier observations for most of Asia are sparse to very sparse [2]. One parameter contributing to the annual mass balance of glaciers is the amount of solid precipitation (snow) and the snow cover extent. Actually, most glaciers worldwide, rely on the input of solid snow to grow glacier ice [3]. To measure and monitor temporal and spatial changes of snow extent, remote sensing technology from space is considered as an optimum tool (e.g., [4]).

In particular, optical systems are commonly applied to map temporal and spatial changes in snow cover extent (SCE) (e.g., [5,6]). The lower limit of the SCE for glaciers or ice sheets is defined as transient snowline, a measure of the extent of the snow-covered area at "any instant, particularly during the ablation season" [7]. However, in mountainous areas, optical images are often of limited suitability due to cloud coverage or illumination. Space-borne synthetic aperture radar (SAR) sensors provide data independent of prevailing weather and illumination.

Dry snow of up to few meters thickness is considered as transparent for SAR data at C-band [8] frequencies. However, in wet snow conditions, the high attenuation characteristics [9] of water and the significantly increased dielectric permittivity of melting snow (e.g., [10]) reduce the backscatter coefficient significantly in comparison to dry snow or snow free conditions. Numerous studies used this effect to monitor the extent of wet snow with SAR data for C-band (e.g., [4,8,9,11–14]). In opposition to optical remote sensing data, which are capable of monitoring the SCE, SAR systems are solely sensitive to the area being covered by wet snow, which restrict data to the wet snow covered area fraction (WSCAF) per glacier. Several studies used the extent of wet snow to derive the transient snowline on glaciers (e.g., [4,12–14]). The SCE has large impacts on the energy balance and, consequently, on the mass balance (B) and runoff of a glacier [8]. However, previous studies (e.g., [12,13]) mentioned that the discrimination of snow and firn for C-band SAR data is impossible for utilizing co-polarized channels. In consequence, these studies failed at monitoring the temporal evolution of the accumulation area ratio (AAR) once the transient snowline retreated above perennial firn areas. Such retreat occurs primarily in years of strong negative mass balances. To overcome this deficit, it is possible to support with optical data (e.g., [6,13]), which involves the named illumination and cloud problems. Utilizing high winter scenes to analyze for retained liquid water in firn as proposed by Brown [15] does not allow for discrimination of snow and perennial firn either. Hence, a robust method to derive the transient snowline from active microwave remote sensing data—which do not have the restrictions of optical sensors—is beneficial for increasing databases of ablation processes for alpine glaciers.

Several authors describe a direct relationship between annual AAR, the equilibrium line altitude (ELA) and B (e.g., summarized by [16]). The prerequisite to establish this relationship are long-term observations of AAR and B for each specific glacier or a general approximation for this relation. Once the relationships have been established and evaluated, reliable AAR estimates from SAR data enable predictions of B solely from remote sensing. Such data help to assess and quantify runoff from glacierized catchments as snow, firn and ice have different surface albedos and, hence, melt rates are varying [17].

This study introduces a two-step workflow for C-band SAR data enabling monitoring of the WSCAF. We analyzed for backscatter distributions of SAR scenes, which were acquired during wet snow conditions over entire elevation ranges. Multi-annual wet snow scenes allowed for correction of topography-related signal effects and the determination of a wet threshold. If areas classified as being wet fall below 50% of the entire glacier area, we further discriminated wet snow and firn by a subsequent threshold. For quality assessment of determined transient snowlines, we used visible and shortwave infrared data from Sentinel-2 (S2), Landsat-7 (L7) and Landsat-8 (L8) missions. Results of the minimum seasonal wet snow extent were used as annual AAR. For Vernagtferner and Hintereisferner, the thus identified relation between AAR and B were compared with the existing field observations.

2. Materials and Methods

2.1. Study Area and Data

For this study, we used SAR data for the Hinteres Ötztal, Tyrol, Austria acquired from January 2015 to October 2018. We included twelve individual glaciers within our analysis, namely: Gepatschferner (GPF), Guslarferner, Hintereisferner (HEF), Hintereiswände, Kesselwandferner (KWF), Rofenberg West and East, Vernagtferner (VF), Vernaglwandferner North and South and Weissseeferner (Figure 1). Details for all individual glaciers are listed in Table 1. All twelve glaciers differ strongly in size, elevation range, slope angle and exposition. Further details for all glaciers can be found in [18]. For simplicity reasons and to prevent subpixel analysis of the SAR data, we summarized all individual glaciers into three areas of interest (AOI) named after the largest glacier per AOI. Both Guslarferners, both Vernaglwandferners, Weissseeferner and Hintereiswände were grouped together with Gepatschferner into the GPF AOI. The HEF AOI consists of both Rofenberg glaciers and Hintereisferner (Figure 1) and VF consists just of Vernagtferner.

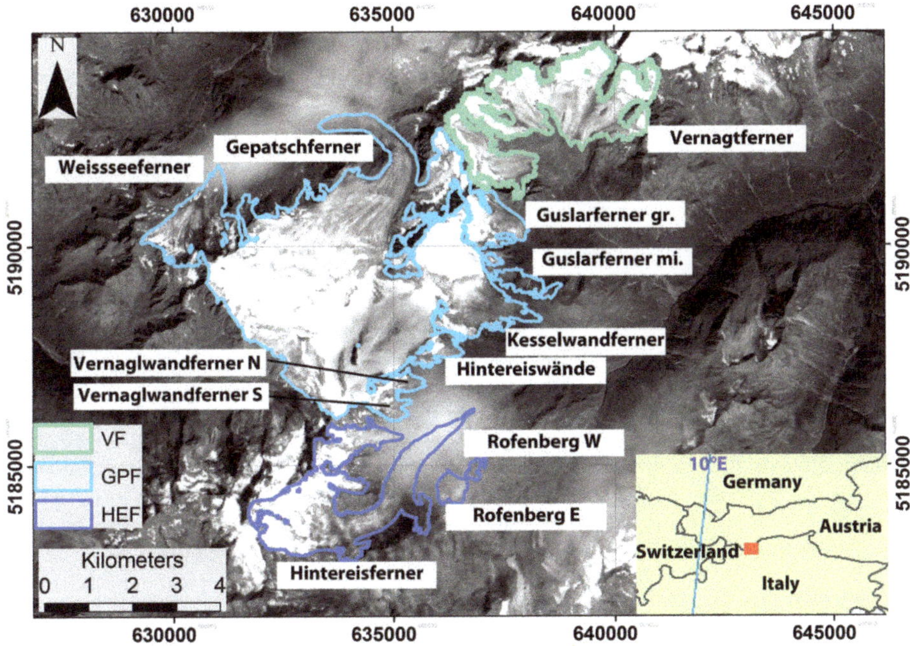

Figure 1. Study area Hinteres Ötztal with named glaciers. Area margins are color coded for the three areas of interest Vernagtferner (VF), Hintereisferner (HEF) and Gepatschferner (GPF). The background image is a Landsat-8 band 8 composite from September 2016. The red rectangle within the inset displays the location of the study area. Coordinates are given in UTM with datum WGS 1984.

Table 1. Glacier names and elevation, exposition as well as size for all individual glaciers observed in this study. Glacier data are taken from [18] with glacier margins from 2006.

Glacier Name	Elevation [m a.s.l.]	Exposition	Size [km^2]
Gepatschferner	2180–3507	NW–NE	16.6
Guslarferner gr.	2842–3479	NE–SE	1.4
Guslarferner mi.	2928–3317	NE	0.5
Hintereisferner	2484–3711	E–NE	7.5
Hintereiswände	3091–3428	SE	0.5
Kesselwandferner	2792–3492	NE–SE	3.8
Rofenberg E	2937–3173	NW	0.1
Rofenberg W	2885–3268	NW	0.4
Vernagelwandferner N	3003–3267	E	0.3
Vernaglwandferner S	2942–3429	SE	0.6
Vernagtferner	2828–3621	SE–W	8.3
Weissseeferner	2608–3502	N–NE	2.6

The remote sensing data basis for this study consists of 82 Sentinel-1A (S1A) and -1B (S1B) scenes (Table A1), which were acquired in the Interferometric Wide Swath Mode (IW) with dual-polarization (VV/VH). Scenes cover an area of 250 km $\times$ 200 km with ground resolution of 10 m $\times$ 10 m. The absolute radiometric accuracy is given by 1 dB [19]. Further details are listed in Table 2. Not every envisaged date of acquisition was achieved in 2015–2018. In addition, right after the launch of S1A and S1B, data acquisitions are usually sparse (Table A1). However, in summer 2017 and 2018, we were able to download almost all theoretically possible S1 scenes with a return cycle of six days. In addition, we collected eight optical and near infrared imagery during the ablation seasons in almost cloud free conditions.

Table 2. Parameters of the acquired Sentinel-1 (S1) data. Values for the noise equivalent sigma zero (NESZ) were derived from [19]. The inclination angle of the respective orbits vary for locations within the areas of interest and consequently are given as approximate values.

Parameters	S1
Band	C
Repeat Pass	12 d, 6 d
Polarization	VV, VH
Orbit Ascending	15, 117 (46°, 37°)
Orbit Descending	168 (39°)
Acquisition Time	05:30 D/ 17:10 A
Acquisition Period	01/2015–10/2018
NESZ	−22 dB
Number Scenes	82

To relate data interpretation from SAR scenes to prevailing meteorological conditions, we used continuous ablation measurements and related meteorological observations from the monitoring program of the Geodesy and Glaciology group of the Bavarian Academy of Sciences, Munich, Germany [20]. On Vernagtferner, at 2930 m a.s.l., a ventilated thermometer records air temperature and an ultrasonic ranger measures ablation and accumulation continuously. For this study, we made use of the hourly data set. We used the ultrasonic data to determine whether new snow per day occurred. In a first step, we calculated the mean ablation rate per season (1 June–1 September each year), which for all four years is strongly negative. Next, we calculated the diurnal trend in surface height from 6:00 UTC to 6:00 UTC the subsequent day and divided each diurnal trend by the respective ablation trend. To relate positive quotients to new snow events, we multiplied by −1 and set all negative trends to zero. This results in a simple solid precipitation index with an indication of increases in surface height per day (new snow event).

To compare mass balance projections with field data, we used the long-term records for mass balances for VF (since 1965) and HEF (since 1952) [21,22]. The glaciological mass balance method relies on stake observations at specific points [23]. Uncertainties for stake readings are rather small (<3–5 cm). However, measurements have to be interpolated in between the stakes. Depending on the number of stakes and interpolation techniques, errors for specific areas on the glaciers can become significantly larger, especially for ice margins or areas difficult or impossible to probe (crevasse zones, etc.). Zemp et al. [24] summarize numerous sources of errors for the the glaciological method and present literature on uncertainties of 100 mm w. e. [25] to 600 mm w.e. [26]. Zemp et al. [1] conclude that systematic errors in glaciological mass balance assessments are usually below 100 mm w.e., while random deviations per year can reach values of "a few hundred mm w.e.".

On 20 September 2018, we identified the transient snowline in situ over a length of about 300 m separating wet snow and firn. This in situ transect represents only a small part of VF. Locations of the transient snowline were recorded with a conventional handheld GPS. We compared this GPS transect with derived snowlines from remote sensing data (SAR and optical results) by intersecting the respective lines and calculating for average offsets.

2.2. SAR Data Workflow

For the complex terrain in the investigated area, it is important to discriminate backscatter distributions for assumed homogeneous surface conditions from backscatter dependencies due to various surface conditions such as wet snow cover, firn cover or bare ice. Under the assumption of dry snow being transparent for SAR data in C-band frequency ranges, the only period of homogeneous glacier surface conditions are after melt affected upper elevation ranges and before the lowermost glacier parts become snow free. Otherwise, contributing surfaces have different dielectric permittivities (dry, wet snow, ice) and highly variable surface roughnesses. For the Alpine region discussed here, complete wet snow coverage occurs usually in June each year followed by an upglacier retreat of the SCE during the summer. In the following, we describe data processing for all acquired SAR scenes including statistical analysis of homogeneous backscatter scenes. The workflow to derive transient snowlines from acquired SAR scenes is displayed in Figure 2.

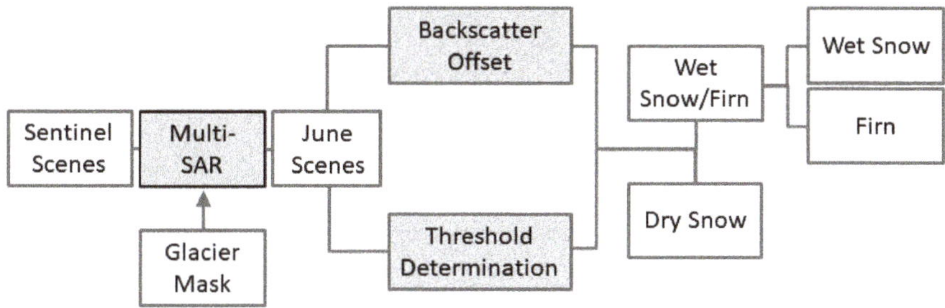

Figure 2. Workflow for the classification of wet and dry snow and firn based on Sentinel-1 data. Grey shaded boxes indicate data processing.

2.2.1. Multi-SAR-System

All acquired SAR data were processed with a pre-processing framework called Multi-SAR-System. The Multi-SAR-System enables processing of SAR data being acquired with different sensors within the same system and output of the processed data in a uniform format. The pre-processing system is implemented at the Earth Observation Center of the German Aerospace Center, Oberpfaffenhofen, Germany. It contains geocoding, radiometric calibration and image enhancement. During geocoding, radar image distortions induced by the side-looking geometry of SAR-systems are corrected using a digital elevation model (DEM). The geometric quality of the geocoding depends on the height accuracy and the resolution of the elevation model. We used an airborne laser scanning DEM with a spatial resolution of 10 m generated between 2006–2010 (data source: Province of Tyrol—data.tirol.gv.at). The next step, after geometric adjustments, was a coarse radiometric calibration. Local topographic variations and sensor position affect not only location, but also the brightness of the radar return [27]. In mountainous terrain, this effect can be reduced with a gamma correction. This approach compares the ratio of individual area parcels of a synthetic image derived from DEM and orbit information with the illuminated SAR images. Due to the integration of backscatter over the area parcels, one S1 scene require about one full day of processing time. For regions with moderate slope angles such as for the glacier areas investigated here, the approximation of the gamma correction based on the local incidence angle using $\gamma^0 = \beta^0 tan\Theta$, γ^0 the backscatter coefficient, β^0 the backscatter in beta naught and Θ the local incidence angle is sufficient. The processing time using the approximated gamma correction reduces to about 20 min per scene. Our interpolation uses the cubic convolution method on a 17×17 pixel raster. The last step of the Multi-SAR-System is the image enhancement to reduce the influence of additive and multiplicative noise contribution. Compared to conventional speckle filtering algorithms, the multi-scale, multi-looking approach applied here adapts the local number of looks to the image content. For heterogeneous surfaces (i.e., mountainous areas), a minimal look number and, consequently, the maximum geometric resolution is necessary to adequately describe surface features. The choice of an appropriate look number is made by the help of a novel perturbation-based noise model that combines both additive and multiplicative noise contributions and automatically adapts to sensor and imaging mode characteristics via the delivered metadata. The result is a very smooth, but detail-preserving multi-looked SAR image representing the local optimal trade-off between geometric resolution and radiometric accuracy. Finally, image values are converted to the standard unit decibel (dB) [28,29].

As a next step, two different masks are applied. The first mask eliminates the radar shadow and layover effects and the second mask removes non-glaciated areas (based on the glacier boundaries of 2009 derived from optical remote sensing data). As a result, we receive SAR data of just the respective glaciated areas described in Figure 1.

2.2.2. Backscatter Variability of Wet Snow Scenes

To determine the variability in γ^0 for homogenous wet snow surfaces, we calculated the coefficient of variation (CV, $CV = \frac{\mu}{\sigma}$, μ the arithmetic mean and σ the standard deviation) per June scene. A low variability in backscatter distribution per glacier areas indicate homogeneous conditions. We set the CV in backscatter to below 20% as criterion for data selection. Table 3 displays the determined CVs for June scenes in cross- and co-polarization.

Almost all cross-polarized SAR scenes in Table 3 show a CV for June acquisitions of 15% or less. The single exception is the S1 scene being recorded in ascending orbit in 2015. All co-polarized scenes, however, result in a variation significantly higher than for cross-polarized acquisitions and show a much larger range. Such larger ranges indicate less consistency. Consequently, we use only on cross-polarized SAR scenes.

A secondary criterion for data selection is time of acquisition. During the main melt season in August, glacier ice and firn surfaces are very much affected by surficial melt water streams during the day, while early in the morning at 5:30 UTC (local time 7:30) nocturnal refreezing has peaked. Unfortunately, all ascending scenes were acquired at 17:10 UTC (local time 19:10), when it is more likely that strong surface wetting and meltwater runoff can lead to misinterpretation for the extent of wet snow covered areas. Hence, in the following, we will present methodology and results only for descending cross-polarized orbits, which were acquired at 5:30 UTC in the morning (Table 2). We excluded the first S1 June scene in Figure 3b (recorded on 2 June 2015) despite a CV value of 14% (Table 3). The median value of this scene is distinctly higher than for the remaining scenes (Figure 3). We expect that higher elevation ranges for all glaciers were still covered by dry snow and, hence, increase the median backscatter to higher values.

Table 3. Determined coefficients of variation for all acquired Sentinel-1 scenes in June between 2015–2018. Respective orbits are described in Table 2.

Platform	Orbit	Date	VH	VV
S1	168	2 June 2015	0.14	0.24
S1	168	8 June 2016	0.10	0.20
S1	168	3 June 2017	0.13	0.25
S1	168	9 June 2017	0.13	0.26
S1	168	15 June 2017	0.14	–
S1	168	21 June 2017	0.15	–
S1	168	4 June 2018	0.13	–
S1	168	10 June 2018	0.14	–
S1	168	16 June 2018	0.15	–
S1	168	28 June 2018	0.15	–
S1	117	10 June 2015	0.13	0.26
S1	117	22 June 2015	0.21	0.44
S1	117	3 June 2015	0.10	0.17
S1	15	15 June 2015	0.11	0.20
S1	15	27 June 2015	0.11	0.18

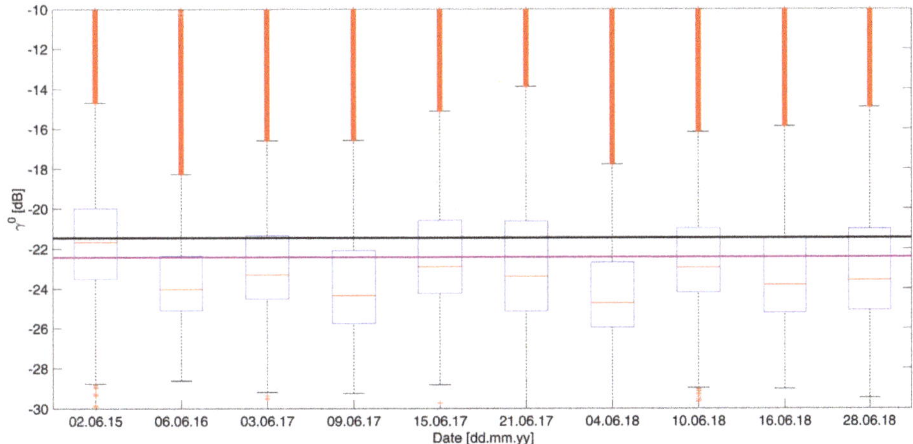

Figure 3. Box plots of backscatter distributions of all recorded cross-polarized June acquisitions in descending orbits with calculated coefficients of variation below 20% in Table 3. The red horizontal lines within the boxes represent the median in γ^0, the boxes frame the interquartile range and the whiskers display extreme values not considered as outliers. Outliers are presented through red crosses. The black horizontal line displays threshold $\beta_1 = -21.45$ dB and the purple line displays threshold $\beta_2 = -22.41$ dB—for details, see Section 2.2.4.

2.2.3. Correction of Systematic Backscatter Offsets

However, cross-polarized wet snow scenes are very homogeneous (Figure 3), and topographic effects very often bias interpretation of satellite data. In addition, very complex topographies, such as mountain glacier regions with various slope aspects, are challenging for space borne acquisitions to interpret. In addition, we relied on a DEM from 2006 for data processing. In particular, at glacier tongues, topography has lowered significantly between time of acquisition and DEM generation. Since we had several acquisitions in June with equal orbits during four years of data acquisition, we could check whether variations to average γ^0 per scene are fully random or very similar for consecutive years and various melt progresses. We calculated for each June SAR scene deviations from the respective median. Deviations in γ^0 being constant in location and independent from melt progress can be attributed as systematic due to topography and incidence angle. Next, we calculated the variance of determined deviations for all June scenes for each pixel. Calculated variances for the nine wet snow scenes below a value of 0.5 indicate that offsets from median are systematic and are corrected for. Figure 4 displays the resulting constant deviations used as backscatter offset correction file. To prevent misinterpretation of snow-free glacier tongues, we set corrections for the GPF and HEF glacier tongues to zero. This file was subtracted from each single S1 scene as additional topography correction. Instead of manually discarding snow-free glacier termini, one can apply automated detection algorithms (e.g., [30]) to minimize influences of already snow free glacier parts. We consider platform effects such as variations in perpendicular baseline or image frame as negligible for the here performed analysis. However, in case such deviations would be significant, the effect per pixel would vary with different baselines and, hence, the pixel variances of different scenes increase.

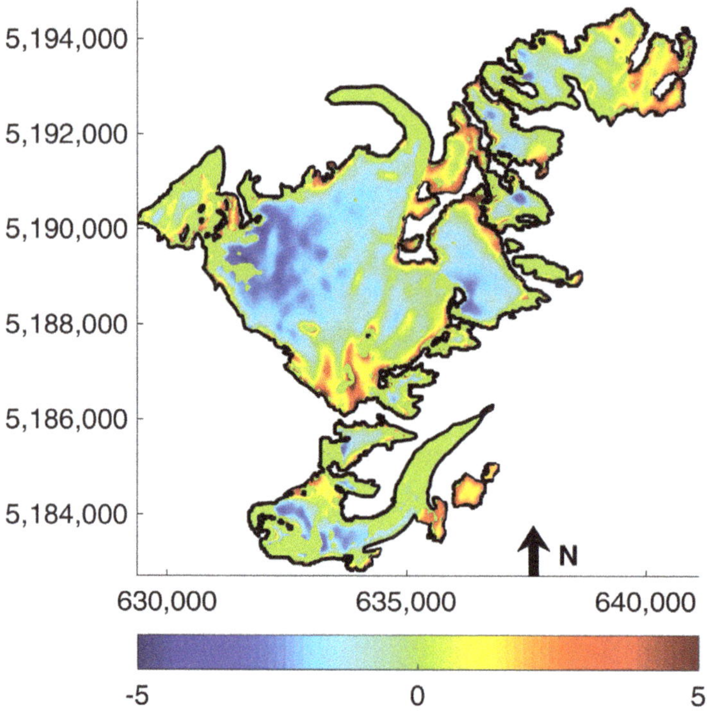

Figure 4. Applied backscatter coefficient (γ^0) correction in decibel (dB), which is subtracted from each scene. Coordinates are given in UTM with datum WGS 1984.

2.2.4. Threshold Determination

The above selected nine wet snow scenes (Figure 3) are used to calculate wet snow thresholds. We calculated for each scene the upper quartile (75% percentile, upper frame of the boxes (Figure 3) of the backscatter distribution. The arithmetic mean of those upper quartiles is used to discriminate wet snow and wet firn from dry snow and bare ice areas (threshold β_1, black line in (Figure 3). In a second step, we used solely the parts of the wet snow scenes with $\gamma^0 < \beta_1$ and calculated the 95% percentiles per remaining part of the scenes. The arithmetic mean of those percentiles result in threshold β_2 (purple line in (Figure 3). The resulting values are $\beta_1 = -21.45$ dB (with standard deviation of 0.76 dB) and $\beta_2 = -22.41$ dB (with standard deviation of 0.18 dB). The determined standard deviations are further used for sensitivity analyses.

For the applied two-step approach, we first analyze for glacier parts being classified as wet using β_1. After correction for systematic backscatter offsets, we subtracted β_1 from each S1 scene resulting in a classification as dry/bare ice (positive values) or wet (negative values). In case the classified wet area is less than 50% of the glacier, we further discriminated for firn and wet snow utilizing β_2. Here, we used again the offset corrected S1 scenes, removed all dry parts classified with β_1 and subtracted the wet areas by β_2. The resulting pixels were classified as wet snow (negative values) and firn (positive values). The proportion of each glacier AOI being either wet or wet snow divided by the whole AOI area results in the WSCAF. Glacier parts, which were covered by dry snow, cannot be distinguished from bare ice areas. As a result, we interpret WSCAF per glacier AOI as accumulation area fraction.

2.3. Processing of Optical and Near-Infrared Remote Sensing Data

Similar to, e.g., Nagler et al. [9] and many other authors, we used L7, L8 and S2 acquisitions to assess the accuracy of the presented SAR workflow on capturing transient snowlines. From now on, we use the term "optical" remote sensing data to abbreviate the usage of Landsat or Sentinel-2 data.

While the SAR data presented here are solely sensitive to wet snow, data within and close to the optical frequency range display brightness differences at the surface. However, cloud coverage often prevents data usage from optical instruments. For instance, in summer 2017, not a single Landsat scene for the here described AOIs could be used for comparison with SAR data. Each Landsat scene for this region was strongly influenced by clouds (a maximum of 20% cloud coverage per glacier was set as cut-off criterion for data usage). For the other observed summer seasons, we were able to make use of eight optical scenes. We used the normalized difference snow index (NDSI) [31–33] for automatic classification of snow covered areas for Landsat data and the Level-2A algorithm for Sentinel-2 data [34]. For L8 scenes, we used a threshold of 0.6 and for L7 this threshold had to be increased to 0.88 to obtain reliable results for snow covered areas. Snow classification thresholds for S2 data were increased to 0.16 (2017) and 0.18 (2018) from the given 0.12 [34] to match best the visually inspected snow boundaries. Since coincidences of optical sensor acquisitions and SAR imagery are rare, the quantification of spatial deviations between both data interpretations is difficult. We defined a range of four days prior and four days after each optical acquisition as being reasonable for direct comparison. For the case of new snow events within this time range, data were not used for analysis.

2.4. Annual Minimum Wet Snow Covered Area Extent

For remote sensing data, fixed date acquisitions such as annual mass balance assessments per glaciological year (GY) are difficult to achieve due to overpass rates and priority rankings by the satellite operators. However, more importantly, conditions on the glacier can be very different from year to year at fixed dates. In this study, we propose searching for local minimum in a wet snow extent within a range of five weeks prior to 30 September each year and two weeks after that day. Within this time range, we search for the local minimum in WSCAF for SAR scenes. However, recent new snow falls shortly before a SAR acquisition influence the detected WSCAF drastically. A minimum by early October could arise due to the fact that almost the entire glacier was covered by dry snow and, hence, volume scattering drastically increased. To prevent misinterpretation of derived snowlines, we applied a correction algorithm before minimum search based on observed firn and snowlines from optical data. We used the fact that perennial firn is more resistant to melt than seasonal snow. For the last analyzed S2 scene (16 September 2018), we decreased the snow threshold to 0.0 (instead of 0.18) and determined the area per AOI for reflectivities larger than zero. These areas include, at least, a fraction of the perennial firn coverage per AOI. Since those firn areas were formed in previous years from snow reserves, it is impossible that the SAR detected firn and WSCAF (result of step 1) is below those optical results. We defined that any SAR scene with lower "wet" areas than 75% of this optical reference scene as being influenced by recent new snow. Hence, those scenes were excluded from annual minimum snowline searches. The resulting minimum snowline is used as annual AAR from SAR scenes.

3. Results

3.1. Wet Snow Covered Area Extent

Figures 5 and 6 display the seasonality of backscatter distributions being above or below the thresholds. For all four observed accumulation seasons (winter periods), snow surface conditions are constantly dry and, hence, WSCAFs are at about zero. Since we determined no changes during periods from January until early April in 2015 and 2016, we minimized data acquisition for this period for the remaining years. From late April to May, the snow surface is starting to becoming wet and reaches towards average WSCAFs in June of 80–90%. June 2015 is an exception, when only 60–70% are reached. However, we could only analyze a single June scene in 2015, which was recorded at the very beginning of the month. After the launch of S1B in spring 2016 and the following commissioning phase, C-band data were available with six days return cycles beginning in late September 2016. The ablation seasons 2017 and 2018 provide a much more detailed view on temporal and spatial changes of WSCAF for all AOIs. WSCAF recession during the ablation season after June is usually very rapid with interruptions by new snow events (Figure 6). New snow events during the ablation season can have a large effect on B [35] due to the short term increase in albedo up to values of 0.9. In summer 2017, two snow fall events can be directly related to sharp increases in WSCAF for all three AOIs. It is possible to detect a peak in WSCAFs for 15 July 2017 for all three AOIs, while respective amplitudes are different. VF decreases after 1 July 2017 from about 80% WSCAF to below 20%, the same occurs for HEF, whereas GPF remains at 60%. The corresponding snow fall events were recorded for 6 and 7 July. However, no increase in WSCAF as a consequence of this recent new snow is recognizable for the subsequent SAR analysis on 9 July 2017. Temperatures remained low and, hence, this recent snow was not affected by melt. Shortly before the following SAR image, snow melt changed γ^0 resulting in a local peak of derived WSCAF. The next peak in WSCAF observed at 20 August 2017 is related to new snow as well. Again, amplitudes are significantly larger for GPF in comparison with VF and HEF. In summer 2018, WSCAFs decrease to very low values for VF and HEF with only slight interruptions. Solely, the new snow event in late August this summer has a remarkable effect on WSCAFs for those two AOIs. GPF, however, is characterized by two additional peaks in July and mid August. At the Ultrasonic location on VF, no snow fall could be recognized. At least for mid August, we noticed a liquid precipitation event at lower elevations, while higher glacier regions received new snow. The mostly northerly exposed GPF, which has a large flat plateau above 3100 m a.s.l., results in a WSCAF increase to above 30% for this SAR scene.

The applied two-step approach has an effect on sensitivity displayed through errorbars. Large errorbars can—in most cases—be attributed to recent new snow precipitations (Figures 5 and 6). After a snow fall event, the area extent of wet snow change across the criterion of 50% for the application of the β_2, which, as a consequence, can have a strong effect on determined WSCAFs. This is regularly the case for GPF.

For comparison with optical remote sensing data, four SAR scenes were available, which fulfill the requirements of a temporal offset of maximum ± 4 days. The root mean square (RMS) deviation for the four periods for VF and GPF and only two periods for HEF (due to cloud coverage) results in 8.5% difference. This deviation is insensitive to subtraction of standard deviations from β_1 and β_2 (RMS deviation of 7.8%) but increases to a RMS deviations of 18.1% when applying the higher threshold range. Unfortunately, a new snow event during the night of 14 to 15 August 2018 above roughly 3200 m a.s.l. had a significant influence on the SAR detected WSCAF at 15 August in the morning. This new snow was already melted before the acquisition of optical imagery the subsequent morning. Discarding this SAR value from RMS analysis leads to deviations of only 4.9% with an sensitivity range of 4.7–13.5%

We used changes in snow/ice height measured at the lower glacier tongue of VF as a proxy to assess ice melt progress until the end of the respective GY. For all three analysed ablation seasons, the depicted annual minimum in snowline from SAR data was before the respective end of the GY. To relate ablation progress, we compare ice heights for these dates with 30 September each year. In 2016, the ice surface lowered by −27 cm from 12 September until 30 September (to −2.66 m). The subsequent year, the offset in ice surface reduced to 19 cm for an even longer time span (26 August–30 September) with an annual ablation of −3.1 m at the end of the GY. Finally, in 2018, we observed a surface lowering of 28 cm for only ten days between SAR minimum and end of GY (−4.53 m). All surface height values are related to the start of the respective GY, which is set to a surface height of 0.0 m. No annual mass balance assessment for GPF are conducted so far.

Offsets to estimates from field investigations are presented in Table 4. RMS deviation result in 8.2% for HEF and VF for a sample size of five.

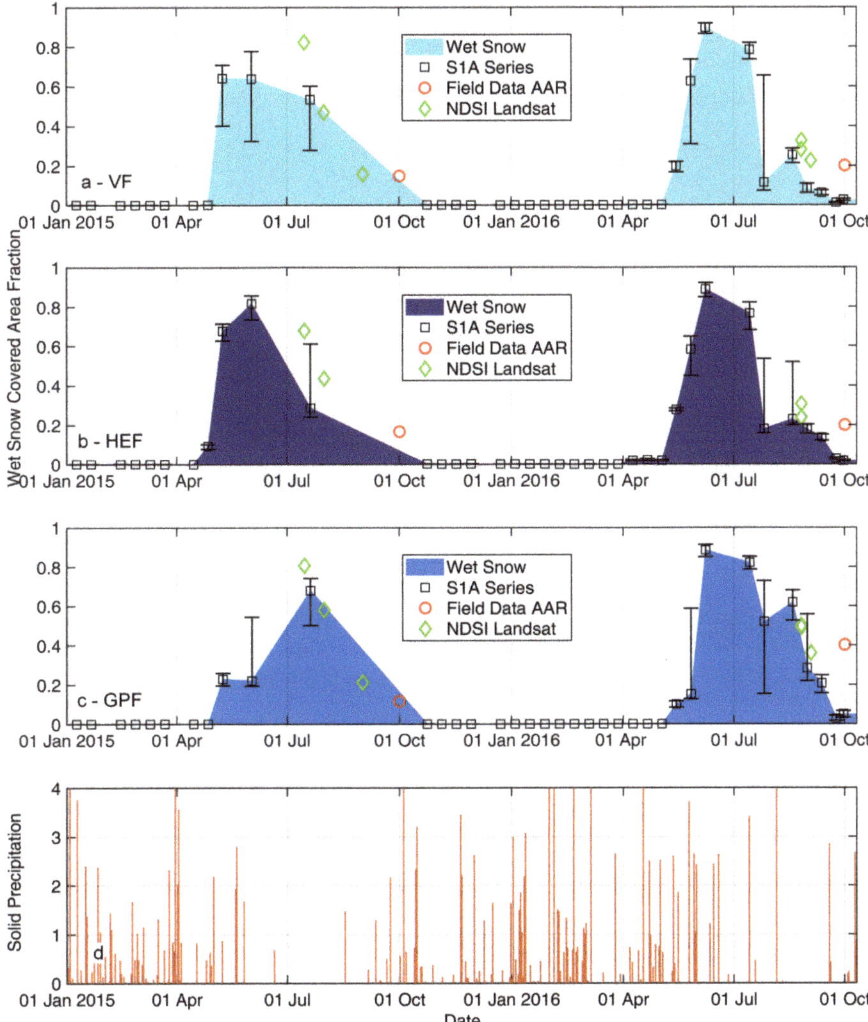

Figure 5. Wet snow covered area fraction for Sentinel-1 (S1) series per area of interest (AOI) Vernagtferner (VF) (**a**), Hintereisferner (HEF) (**b**), Gepatschferner (GPF) (**c**) and results for the new snow index (**d**) from January 2015–October 2016. Black rectangles display S1 scenes with error bars for uncertainties in thresholds, red circles show field data determined accumulation area ratio (AAR) and green diamonds display results for the normalized difference snow index (NDSI) from Landsat images and snow classification results from Sentinel-2 data. Field data for the AOI GPF in (**c**) were only acquired for Kesselwandferner.

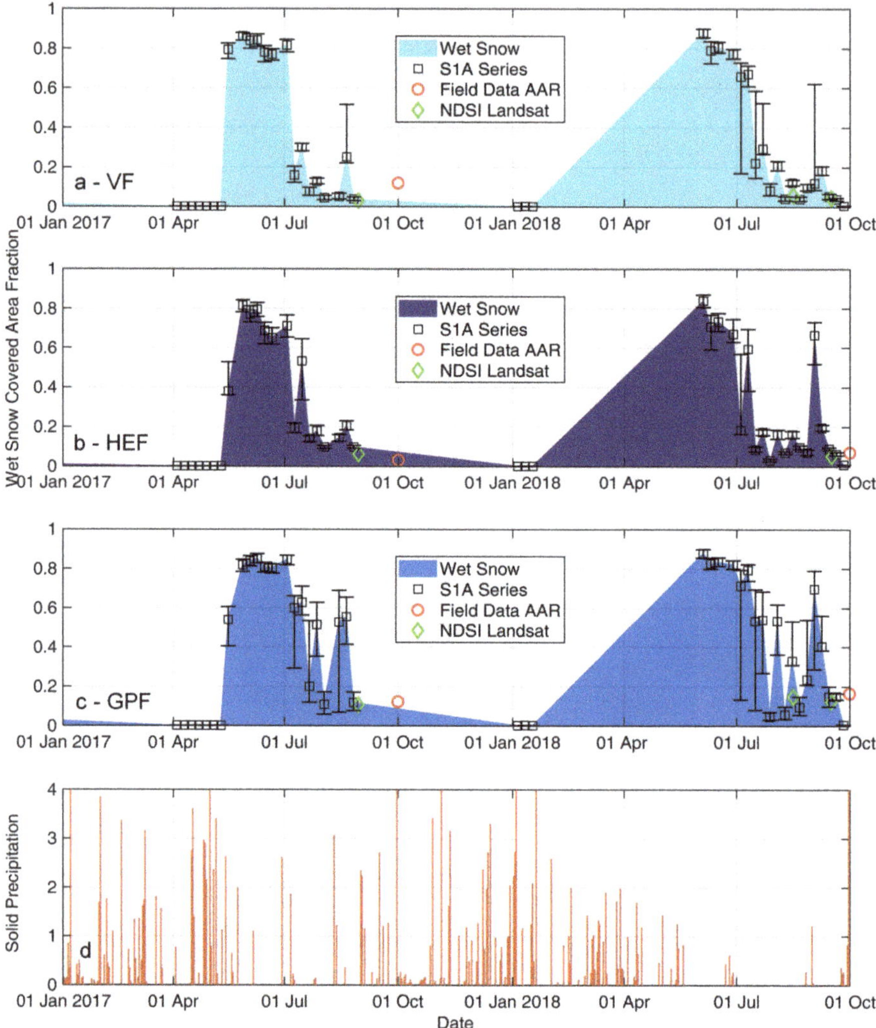

Figure 6. Wet snow covered area fraction for Sentinel-1 (S1) series per area of interest (AOI) Vernagtferner (VF) (**a**), Hintereisferner (HEF) (**b**), Gepatschferner (GPF) (**c**) and results for the new snow index (**d**) from January 2017–October 2018. Black rectangles display S1 scenes with error bars for uncertainties in thresholds, red circles show field data determined accumulation area ratio (AAR) and green diamonds display results for the normalized difference snow index (NDSI) from Landsat images and snow classification results from Sentinel-2 data. Field data for the AOI GPF in (**c**) were only acquired for Kesselwandferner.

Table 4. Annual accumulation area ratio (AAR) estimates from field data and annual minimum in wet snow covered area fraction determined from Sentinel-1 (S1) scenes with date of acquisition in brackets per glaciological year (GY) and area of interest Vernagtferner (VF) and Hintereisferner (HF). All values are presented in percent (%).

		VF		HEF
GY	**Field Data**	**S1**	**Field Data**	**S1**
2015/16	20.0	6.2 (12 Sep.)	20.0	13.7
2016/17	12.0	3.9 (26 Aug.)	3.2	9.7
2017/18	–	3.9 (20 Sep.)	7.1	6.2

3.2. Discriminating Firn and Wet Snow

Only reliable discriminations between wet snow and firn enable derivation of annual AAR for various glaciers. We analyze for accuracies in wet snow–firn discrimination by a short in situ GPS transect for the transient snowline on VF. For this 303 m long transect discriminating firn from wet snow, we determined an average offset for the SAR-derived transient snowline of 35 m and for optical data this offset decreases to 19 m. In addition, 35 m correspond to about 3–4 pixels for a 10 m × 10 m SAR resolution. The respective S2 data providing optical data has the same ground resolution. However, optical data have been recorded four days prior to ground truth and S1 data. Apart from the short 300 m in situ line, qualitatively, we demonstrate agreement of optical data with SAR data in Figure 7. It is inevitable that disagreement between optical data and SAR data occur, but, in general, this overview confirms the 3–4 pixel accuracy determined for the in situ data.

3.3. From Minimum Wet Snow Extent to Mass Balances

For Vernagtferner and Hintereisferner, long-term summer and winter mass balance series exist (VF: 1964/65–2016/17; HEF: 1952/53–2017/18; [21,22]) (e.g., [36]). We plotted relationships of AAR and mass balance (B) for VF and HEF in Figure 8. It is clearly visible that a linear fit matches the relation between B and AAR adequately. Coefficients of determination (R^2) are high reaching $R^2 = 0.93$ for HEF with a sample size of 66 years and a range of 0% to above 80% in AAR. The $R^2 = 0.90$ for VF is slightly lower with a lower sample size of 53 years and a higher range from 0% to more than 90% in AAR. Calculated RMS deviations to the linear approximation for both glaciers are at 212 mm w.e. for VF and 154 mm w.e. for HEF.

Such long-term mass balance series with reliable relationships of AAR and B enable direct conversion from SAR determined annual AAR to B (Figure 8, Table 5). Deviations to observed B values from field data are highly variable from +435 mm w.e. to −73 mm w.e. for VF and from +248 mm w.e. to −241 mm w.e. for HEF using S1 data. In absolute deviations, those numbers average to 254 mm w.e. for VF and 231 mm w.e. for HEF. Such average offsets are above the given accuracy ranges of the linear approximation (in RMS deviation); however, sample numbers are very low.

Instead of an individually derived linear relationship for each single glacier, Dyurgerov et al. [37] used 99 index glaciers and came up with an average relationship for AAR and B. We included results from this average formula in Table 5 and Figure 8, together with a relationship derived from just Eastern Alpine glaciers (11 glaciers with B and AAR values) listed in [37]. For both glacier areas and most observation years, offsets in mass balance values surpass the given uncertainty range for direct measurements with the glaciological method significantly. Dyurgerov's approximation result in an average absolute offset for VF of 797 mm w.e.. However, the rather typically shaped valley glacier area of HEF matches the general approach by Dyurgerov significantly better. Average absolute offsets sum up to 190 mm w.e.. An approximation established just for Eastern Alpine glaciers does not decrease offsets. Calculated average absolute deviations are 377 mm w.e. for VF and 367 mm w.e. for HEF. The sample number for the Eastern Alps is very low with just 11 glaciers; however, the R^2 of the linear approximation reaches 0.79 and the RMS deviation for the 11 glaciers results in 179 mm w.e.

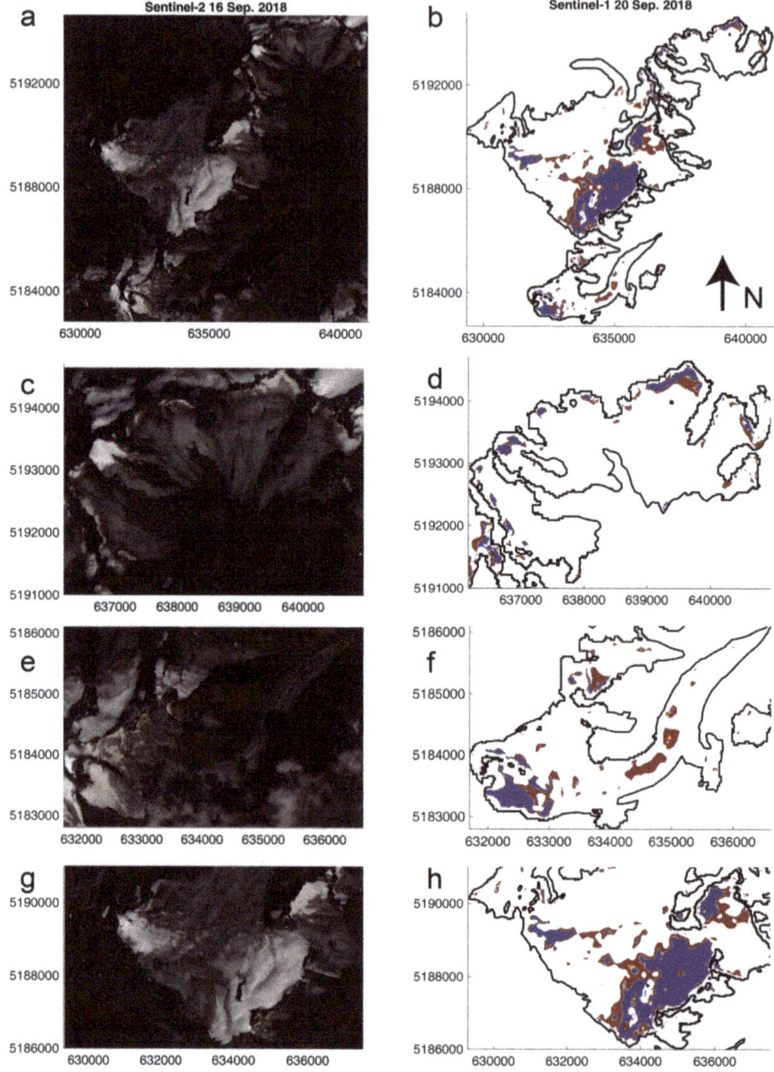

Figure 7. Comparison of Sentinel-2 (S2) optical data showing firn and snow patches with Sentinel-1 (S1) derived wet snow covered area fraction maps for all areas of interest (AOIs) and for each AOI respectively. The color coding is constantly displaying derived firn (red) and wet snow (blue) areas for all presented S1 data. S2 imagery was recorded on 16 September 2018 and S1 data on 20 September 2018. S2 imagery (**a**), S1 analysis presenting all three AOIs (**b**), zoom for AOI Vernagtferner with optical (**c**) and S1 data (**d**), zoom for AOI Hintereisferner with optical (**e**) and S1 data (**f**), zoom for a fraction of AOI Gepatschferner with optical (**g**) and S1 data (**h**). Coordinates are given in UTM with datum WGS 1984. All maps are aligned equally.

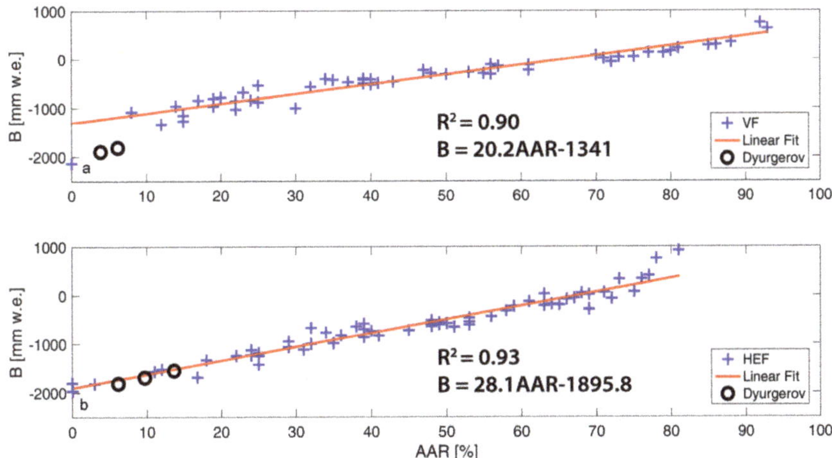

Figure 8. Mass balance (B) versus accumulation area ratio (AAR) distribution for 53 years of records for Vernagtferner (VF) (**a**) and 66 years of records for Hintereisferner (**b**) (HEF). Black circles indicate results from the linear approximation by Dyurgerov et al. [37]. Correlation coefficients and formulation for the linear fit are presented.

Table 5. Annual mass balance (B) estimates from field data and annual minimum accumulation area ratios (AAR) determined from Sentinel-1 (S1) scenes. For conversion from accumulation area ratio (AAR) to B, we used the formulations of the linear fit for Vernagtferner (VF) and Hintereisferner (HEF) displayed in Figure 8, Dyurgerov's average relationship [37] (Dyu) for all glaciers and specified just for the Eastern Alps (Dyu EAlps). All mass balance values are presented in millimeter water equivalent (mm w.e.).

	VF				HEF			
GY	Field Data	Linear Fit	Dyu	Dyu EAlps	Field Data	Linear Fit	Dyu	Dyu EAlps
2015/16	−781	−1216	−1814	−1406	−1263	−1511	−1547	−1223
2016/17	−1335	−1262	−1896	−1463	−1826	−1623	−1689	−1321
2017/18	–	−1262	−1896	−1463	−1963	−1722	−1814	−1406

4. Discussion

4.1. Uncertainty Analysis and Limitations

All presented data on AAR and mass balances include uncertainties. It is difficult to quantitatively assess the respective error per observation technique. The assessment of annual AAR is usually more subject to uncertainties. For a glacier-wide estimate of AAR, peaks above the glacier or airplane observations are beneficial to prevent misclassification of recent and perennial firn. For very small accumulation areas left after a strong ablation summer, correct assessments depend on accumulation stakes and direct observations. This is especially the case for a strong negative year in terms of mass balance subsequent to a rather less negative year. Extrapolation for the whole glacier area and interpolation in between observations is challenging.

The introduced two-step thresholding based on wet snow scenes (Section 2.2.4) does include some local uncertainties. We observed areas that showed γ^0 values below the respective threshold, which certainly were snow free in late August and September (Figure 7f). Such areas usually were present on HEF and GPF and, for both glaciers, limited to the glacier tongue regions. It appears from comparing images with and without applied correction for systematic backscatter offsets (Section 2.2.3) that these areas deviate strongly from median γ^0 (Figure 4). The applied backscatter enhancement was probably not sufficient to correct for these deviations. In general, rough surfaces like bare glacier ice increase the reflected backscatter [38] for SAR data due to enhanced diffusive scattering. We can only speculate about the reasons of this amplification of the decline in γ^0. Although the applied DEM does not indicate layover and shadowing for those areas, we assume that the steep topography has some effects, especially because the DEM used for SAR processing was generated up to more than ten years prior to acquisition dates and the glacier surface lowered significantly within this time period.

In general, the presented RMS deviations of WSCAF to results from optical remote sensing imagery are below 10%. However, sample sizes for comparison are low. We only found four optical scenes, which were recorded shortly prior or after an S1 data acquisition. For the three analyzed AOIs, the calculated RMS values represent not very resilient statistical values. An accuracy of below 10% is encouraging, however, considering the fact that conventional methods rely on data inter- and extrapolation or are hampered by cloud coverage. Since for three consecutive years only very few optical scenes were available during the ablation periods, SAR data offer a significant progress in determination of annual and seasonal transient snowlines and, consequently, AAR. Six days of overpass rates enabled already almost complete coverage for the ablation seasons 2017 and 2018 with a high temporal resolution.

In addition, such a high temporal resolution demonstrates difficulties in data interpretation by simple thresholding. The sensitivity of WSCAF results on deviations in thresholds is in most cases very low. Derived uncertainty ranges are at ±2%, which corresponds to the sizes of data markers in Figures 5 and 6. However, some errorbars cross the set 50% WSCAF criterion for different thresholding and, consequently, sensitivity analysis are performed over more than 1 dB. Especially for GPF with its flat plateau above 3100 m a.s.l., fluctuation in thresholding occurs more frequently. In most cases, sensitivity increases shortly after a recent new snow event. New snow has two different effects on WSCAF classification from SAR data. If the recent new snow coverage experiences melt, the WSCAF will significantly increase. In addition, dry new snow increases volume scattering from the surface, and, consequently, resulting backscatter values of former wet surfaces may reach above the set threshold values. Hence, determined WSCAF decreases. Such occurrences can be observed for instance at the end of each GY or shortly after 01 October. Here, WSCAFs reduce to roughly 0%. Disregarding this effect would result in misclassification of minimal wet snow extents per ablation season. The applied criterion for the search of minimal extents in wet snow each year seems to work properly to neglect such dry snow covered minimum in WSCAF.

Since large large threshold sensitivities—displayed via errorbars—can be related to new snow events or WSCAFs at about 50%, such values are prone to errors and uncertainties. For comparison with optical remote sensing data or for determination of annual AAR values, we recommend focusing on SAR data being insensitive to threshold variations. For S1 data with six days return cycles, this would reduce seasonal coverage by one (2017) to four (2018) scenes for VF and limit the temporal coverage for GPF to only half of the previously analyzed scenes.

4.2. Discrimination of Wet Snow and Firn

Not only do dielectric permittivities change backscatter emissivity of surfaces, but surface roughness as well (e.g., [38]). Wet snow and wet firn have similar dielectric permittivities, if the volumetric liquid water content (θ_w) is equal and impurities are negligible (s = snow, f = firn, $\rho_s = 360$ kg/m^3, $\theta_w = 0.04$, $\rho_f = 500$ kg/m^3; $\epsilon_s = 2.8$, $\epsilon_f = 3.2$). Concave furrows as a result of melt and sublimation processes on the snow surface increase with continuing ablation, especially, for the inner Alpine dry regions with intensified sublimation observed here. According to the Rayleigh criterion, surfaces are considered as rough for SAR data, if $h > \lambda/8cos\delta$ [39] with h the height of the surface features, δ the given incidence angles and λ the wavelength of the respective platform. This leads to a roughness sensitivity of roughly $h > 8$ mm for C-band data. For soil surfaces, [40,41] found that SAR backscatter increases are most sensitive to changes in surface roughness of $0 < Zs \leq 0.4$ with $Zs = s^2/l$, the roughness parameter where s is the root mean square of surface height h and l the correlation length. For reasonable s and l values, we derive Zs as displayed in Figure 9. It is obvious that backscatter increases are caused by increases in s between 0 cm to about 7 cm for correlation lengths observed in the field. An increase in surface roughness up to 7 cm can usually be observed for transitions from seasonal snow to firn (see Figure A1). For the short in situ transect separating firn and wet snow recorded the same day as a SAR acquisition, we found sufficient agreement of 3–4 pixels. However, the difference of the thresholds separating firn and snow is close to the given resolution limits of S1 data.

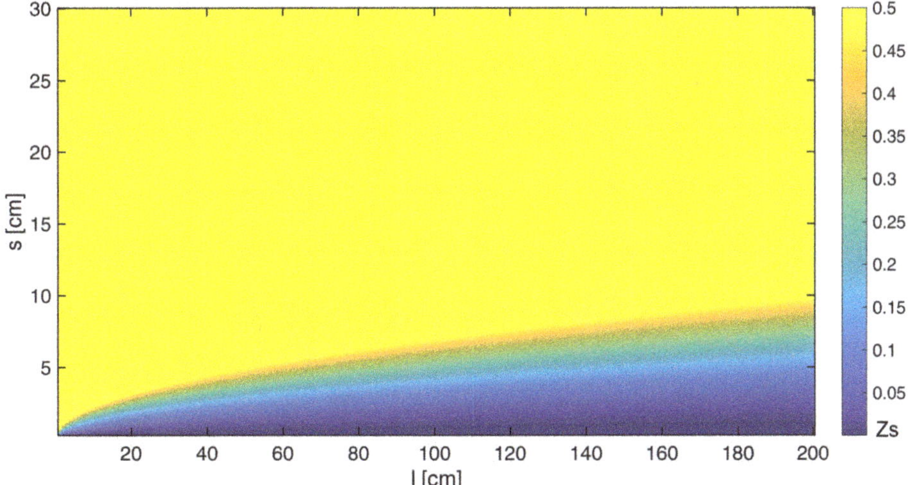

Figure 9. Numerical simulation of variations of the roughness parameter Zs with changes in roughness height s and correlation length l.

4.3. SAR Data for Application in Seasonal Mass Balance Estimates

It remains debatable to which degree field assessments of AAR are accurate. Usually, no continuous snowline data are available for field assessments. Accumulation area ratios for alpine glaciers are commonly estimated according to elevation ranges and stake readings. However, without a long-term time series of direct measurements, no reliable AAR to B relationship can be established.

The long time series of mass balance observations for HEF and VF with data distributions from 0% to almost 100% in AAR enable for deriving a resilient formulation for the AAR to B correlation. Inserting SAR based annual AAR in a linear fit results in average offsets of 200–300 mm w.e. This is above given uncertainty ranges for systematic errors of direct mass balance measurements and above the uncertainty of the linear approximation for each glacier. Two limiting factors reduce accuracy of such an approach: (i) for both time series, the linear relationship resulted in average errors larger than 100 mm w.e. (RMS deviation VF 212 mm w.e., HEF 154 mm w.e.); and (ii) the date of SAR acquisition is crucial. The approach by Dyurgerov et al. [37] with a generalized relationship over 99 glaciers is not sufficient for the rather flat and southerly exposed VF to accurately determine B just from AAR observations. For HEF, as a more typical valley glacier, this generalized approximation provides better estimates on B for the three observed GYs. Reducing the globally derived glacier relationship of B and AAR to the geographical region of the Eastern Alps provides larger deviations for HEF and decreases the offset for VF. However, data used in [37] have a temporal extent until early 2000. Including the strong ablation seasons within the last decade most likely will influence the derived relationship of AAR and B.

Observation of the temporal evolution of the transient snowline with SAR data is reliably possible with the presented algorithm for the glaciers of the Rofental, Austria. Single scenes are very difficult to interpret, since the respective acquisition can be biased by recent new snow precipitation. However, data interpretation benefit from time series of SAR observations with high temporal resolution and thus misinterpretation due to new snow events is limited. In addition, the applied criteria for the search of annual minimum reliably exclude scenes covered by recent dry snow.

Whether the proposed two-step approach for discriminating wet snow and firn is applicable for other areas in various latitudes or climatological regions remains unknown. For glaciers outside middle latitudes without pronounced seasonality (e.g., high latitudes and very high altitudes), remote sensing data only provide snapshots depending on the time of acquisition and prevailing weather conditions. The derived data on WSCAF for such geographical regions has no or limited significance on ablation progress. As a first step for each region, γ^0 values for wet snow conditions have to be analyzed for all acquired orbits and polarization channels. In case sufficient amounts of SAR scenes fully covered by wet snow are available, the processing routines as described here can be implemented into automatic data analysis resulting in an output of just WSCAF per acquisition.

However, such data analysis does not directly lead to surface mass balance per glacier area. The relationship of annual AAR to B is usually significantly different for each glacier and relies on individual topography, elevation range and aspect. Empirical approaches such as the one proposed by Drolon et al. [42] are less influenced by individual glacier topographies and do not need long-term data series. However, [42] focuses rather on large scale accuracies than on individual glaciers and use optical satellite data with the named restrictions on visibility. In addition, presented uncertainties on annual mass balance values are rather large.

Other SAR platforms using X- or L-band sensors are most likely applicable for such an analysis as well (see [43]). Again, for proposed workflow described here, it is prerequisite that wet snow scenes can be used to derive threshold values and to apply topographic corrections. The main disadvantage for those platforms is the temporal resolution with return cycles of more than 10 days. For instance, over several years of ALOS-PALSAR-1 and 2 data, we could only acquire three June scenes for the region analyzed here.

5. Conclusions

This case study presents a two-step workflow using thresholds to discriminate wet snow and firn for SAR data. In a first step, the workflow distinguishes dry and wet surfaces and subsequently analyzes wet surfaces for wet snow and firn with a second threshold value. Deviations to the extent of snow covered areas derived from visual and shortwave infrared channels are less than 10% (8.5%) in area for C-band data. WSCAF is sensitive to increases in threshold values, which can lead to offsets up to 18.1%. Average offsets of annual minimum in WSCAF from SAR data and field data on annual AAR are at below 10% (8.2%) as well for three observed GYs and three different glacier areas. For analysis, we included 12 individual glaciers covering a wide range of elevation and exposition as well as area extent. All glaciers are located within the Rofental, Austria. For two glaciers within this study, long-term mass balance series are available. We used these time series to establish linear relationships between AAR and B and compared field measurements on B with outcomes for the relationships for SAR derived AAR values. Deviations between field assessments of B and SAR derived values are at 200–300 mm w.e. on average. Such results are not an improvement compared with conventional measurements but enable mass balance estimates for regions not accessible or too dangerous for conventional mass balance observations. Results are very encouraging and demonstrate the feasibility of the proposed workflow to derive AAR from SAR data. In addition, we could show that glacier surface conditions during the melt season can be quasi-continuously be monitored using S1 SAR data, which is essential for glacier runoff modeling. Implementation of this workflow for other regions worldwide has to be analyzed with extended SAR data acquisitions. However, apart from SAR data, solely the availability of DEMs with sufficient accuracy is required to apply the proposed workflow.

Author Contributions: A.H. performed the scientific analysis, acquired the optical data, designed the structure and prepared the manuscript. A.W. is responsible for the download of SAR data and geocoding in the Kennaugh processing chain. A.S. provided the Kennaugh processing chain. C.M. contributed with the design and planning of the study and helped to improve the quality of the manuscript. All authors contributed to editing the manuscript.

Funding: This research received funding by the German Science Foundation through DFG grant HE7501/1.

Acknowledgments: We would like to thank the Copernicus Service for providing the Sentinel-1 and Sentinel-2 data free of charge and sincerely appreciate the support by Rainer Prinz, Irmgard Juen and the Institute of Atmospheric and Cryospheric Sciences at University Innsbruck, Austria for providing data and information about HEF and KWF. Landsat data were obtained through GLOVIS USGS. A.H. was supported by DFG grant HE7501/1.

Conflicts of Interest: The authors declare no conflict of interest.

Abbreviations

The following abbreviations are used in this manuscript:

AAR	Accumulation Area Ratio
AOI	Area Of Interest
B	Mass Balance
CV	Coefficient of Variation
DEM	Digital Elevation Model
GPF	AOI Gepatschferner
GY	Glaciological Year
HEF	AOI Hintereisferner
NDSI	Normalized Difference Snow Index
NESZ	Noise Equivalent sigma-0
RMS	Root Mean Square
S1	Sentinel-1
S1A	Sentinel-1A
SAR	Synthetic Aperture Radar
SCE	Snow Cover Extent
SRTM	Shuttle Radar Topography Mission
USGS	United States Geological Survey
UTC	Universal Time Coordinated
UTM	Universal Transverse Mercator
VF	AOI Vernagtferner
VH	Vertical polarisation transmit and Horizontal polarisation receive
VV	Vertical polarisation transmit and Vertical polarisation receive
WGS	World Geodetic System
WSCAF	Wet Snow Covered Area Fraction

Appendix A

Table A1. Date of all analyzed Sentinel-1 (S1) scenes for this study.

S1	S1	S1	S1
9 January 2015	21 January 2015	14 February 2015	26 February 2015
10 March 2015	22 March 2015	15 April 2015	27 April 2015
9 May 2015	21 May 2015	02 June 2015	20 July 2015
24 October 2015	05 November 2015	17 November 2015	29 November 2015
23 December 2015	04 January 2016	16 January 2016	28 January 2016
9 February 2016	21 February 2016	04 March 2016	16 March 2016
28 March 2016	09 April 2016	21 April 2016	03 May 2016
15 May 2016	27 May 2016	08 June 2016	14 July 2016

Table A1. *Cont.*

S1	S1	S1	S1
26 July 2016	19 August 2016	31 August 2016	12 September 2016
24 September 2016	30 September 2016	4 April 2017	10 April 2017
16 April 2017	22 April 2017	28 April 2017	4 May 2017
10 May 2017	16 May 2017	28 May 2017	3 June 2017
9 June 2017	15 June 2017	21 June 2017	3 July 2017
9 July 2017	15 July 2017	21 June 2017	27 July 2017
2 August 2017	14 August 2017	20 August 2017	26 August 2017
5 January 2018	11 January 2018	17 January 2018	4 June 2018
10 June 2018	16 June 2018	28 June 2018	4 July 2018
10 July 2018	16 July 2018	22 July 2018	28 July 2018
3 August 2018	09 August 2018	15 August 2018	21 August 2018
27 August 2018	2 Sept. 2018	8 September 2018	14 September 2018
20 September 2018	26 September 2018	2 October 2018	

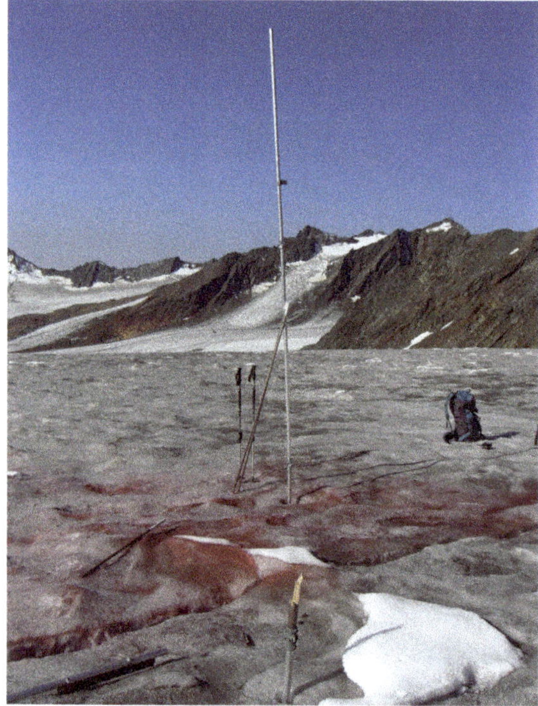

Figure A1. Picture of snow and perennial firn on Vernagtferner recorded in September 2018. The pole in the center is an accumulation stake, which used to be an ablation stake in summer 2018. See the backpack and ski poles for roughness height estimates. The red surface color has been spread out in previous years to detect the previous fall surface for winter field campaigns.

References

1. Zemp, M.; Thibert, E.; Huss, M.; Stumm, D.; Rolstad Denby, C.; Nuth, C.; Nussbaumer, S.U.; Moholdt, G.; Mercer, A.; Mayer, C.; et al. Reanalysing glacier mass balance measurement series. *Cryosphere* **2013**, *7*, 1227–1245. [CrossRef]

2. WGMS. *Global Glacier Change Bulletin No. 1 (2012–2013)*; ICSU(WDS)/IUGG(IACS)/UNEP/UNESCO/WMO; World Glacier Monitoring Service: Zurich, Switzerland, 2015.

3.   Benn, D.I.; Evans, D.J.A. *Glaciers & Glaciation*; Arnold: London, UK, 2006.

4.   Nagler, T.; Rott, H. Retrieval of wet snow by means of multitemporal SAR data. *IEEE Trans. Geosci. Remote Sens.* **2000**, *38*, 754–765. [CrossRef]

5.   Parajka, J.; Blöschl, G. The value of MODIS snow cover data in validating and calibrating conceptual hydrologic models. *J. Hydrol.* **2008**, *358*, 240–258. [CrossRef]

6.   de Ruyter Wildt, M.S.; Oerlemans, J.; Björnsson, H. A method for monitoring glacier mass balance using satellite albedo measurements: Application to Vatnajökull, Iceland. *J. Glaciol.* **2002**, *48*, 267–278. [CrossRef]

7.   Cogley, J.G.; Hock, R.; Rasmussen, L.A.; Arendt, A.A.; Bauder, A.; Braithwaite, R.J.; Jansson, P.; Kaser, G.; Möller, M.; Nicholson, L.; et al. *Glossary of Glacier Mass Balance and Related Terms. IHP-VII Technical Documents in Hydrology No. 86*; IACS Contribution No. 2; UNESCO-IHP: Paris, France, 2011; p. 114.

8.   Adam, S.; Pietroniro, A.; Brugman, M. Glacier snow line mapping using ERS-1 SAR imagery. *Remote Sens. Environ.* **1997**, *61*, 46–54. [CrossRef]

9.   Nagler, T.; Rott, H.; Ripper, E.; Bippus, G.; Hetzenecker, M. Advancements for Snowmelt Monitoring by Means of Sentinel-1 SAR. *Remote Sens.* **2016**, *8*, 348. [CrossRef]

10.   Heilig, A.; Mitterer, C.; Schmid, L.; Wever, N.; Schweizer, J.; Marshall, H.P.; Eisen, O. Seasonal and diurnal cycles of liquid water in snow-Measurements and modeling. *J. Geophys. Res. Earth Surf.* **2015**, *120*, 2139–2154. [CrossRef]

11.   Baghdadi, N.; Gauthier, Y.; Bernier, M. Capability of Multitemporal ERS-1 SAR Data for Wet-Snow Mapping. *Remote Sens. Environ.* **1997**, *60*, 174–186. [CrossRef]

12.   König, M.; Winther, J.G.; Isaksson, E. Measuring snow and glacier ice properties from satellite. *Rev. Geophys.* **2001**, *39*, 1–27. [CrossRef]

13.   Jaenicke, J.; Mayer, C.; Scharrer, K.; Münzer, U.; Gudmundsson, Á. The use of remote-sensing data for mass-balance studies at Mýrdalsjökull ice cap, Iceland. *J. Glaciol.* **2006**, *52*, 565–573. [CrossRef]

14.   Huang, L.; Li, Z.; Tian, B.s.; Chen, Q.; Liu, J.L.; Zhang, R. Classification and snow line detection for glacial areas using the polarimetric SAR image. *Remote Sens. Environ.* **2011**, *115*, 1721–1732. [CrossRef]

15.   Brown, I.A. Synthetic Aperture Radar Measurements of a Retreating Firn Line on a Temperate Icecap. *IEEE J. Sel. Top. Appl. Earth Observ. Remote Sens.* **2012**, *5*, 153–160. [CrossRef]

16.   Barry, R.G.; Gan, T.Y. *The Global Cryosphere: Past, Present, and Future*; Cambridge University Press: Cambridge, UK, 2011.

17.   Huss, M.; Sold, L.; Hoelzle, M.; Stokvis, M.; Salzmann, N.; Farinotti, D.; Zemp, M. Towards remote monitoring of sub-seasonal glacier mass balance. *Ann. Glaciol.* **2013**, *54*, 75–83. [CrossRef]

18.   Lambrecht, A.; Kuhn, M. Glacier changes in the Austrian Alps during the last three decades, derived from the new Austrian glacier inventory. *Ann. Glaciol.* **2007**, *46*, 177–184. [CrossRef]

19.   Sentinel-1Team. *Sentinel-1 User Handbook*; European Space Agency: Frascati, Italy, 2013.

20.   Braun, L.; Reinwarth, O.; Weber, M. Der Vernagtferner als Objekt der Gletscherforschung. *Zeitschrift f. Gletscherkunde und Glazialgeologie* **2011**, *45/46*, 85–104.

21.   Geodesy.; Glaciology. Kenngrößen des Massenhaushaltes des Vernagtferners für den Zeitraum 1964 bis 2015. Available online: http://glaziologie.de/massbal/index.html (accessed on 9 January 2017).

22.   WGMS. WGMS Fluctuations of Glaciers Browser. Available online: http://wgms.ch/fogbrowser/ (accessed on 9 January 2017).

23.   Oestrem, G.; Stanley, A. *Glacier Mass Balance Measurements: A Manual for Field and Office Work*; NHRI Science Report 4; National Hydrology Research Institute: Ottawa, ON, Canada, 1969.

24.   Zemp, M.; Hoelzle, M.; Haeberli, W. Six decades of glacier mass-balance observations: A review of the worldwide monitoring network. *Ann. Glaciol.* **2009**, *50*, 101–111. [CrossRef]

25.   Jansson, P. Effect of Uncertainties in Measured Variables on the Calculated Mass Balance of storglaciaren. *Geogr. Ann. Ser. A Phys. Geogr.* **1999**, *81*, 633–642. [CrossRef]

26.   Funk, M.; Morelli, R.; Stahel, W. Mass balance of Griesgletscher 1961–1994: Different methods of determination. *Zeitschrift f. Gletscherkunde und Glazialgeologie* **1997**, *33*, 41–55.

27.   Small, D. Flattening Gamma: Radiometric Terrain Correction for SAR Imagery. *IEEE Trans. Geosci. Remote Sens.* **2011**, *49*, 3081–3093. [CrossRef]

28.   Schmitt, A.; Wendleder, A.; Hinz, S. The Kennaugh element framework for multi-scale, multi-polarized, multi-temporal and multi-frequency SAR image preparation. *ISPRS J. Photogramm. Remote Sens.* **2015**, *102*, 122–139. [CrossRef]

29. Bertram, A.; Wendleder, A.; Schmitt, A.; Huber, M. Long-term Monitoring of water dynamics in the Sahel region using the Multi-SAR-System. *Int. Arch. Photogramm. Remote Sens. Spat. Inf. Sci.* **2016**, *XLI*, 313–320. [CrossRef]

30. Otsu, N. A Threshold Selection Method from Gray-Level Histograms. *IEEE Trans. Syst. Man Cybern.* **1979**, *9*, 62–66. [CrossRef]

31. Dozier, J. Spectral signature of alpine snow cover from the landsat thematic mapper. *Remote Sens. Environ.* **1989**, *28*, 9–22. [CrossRef]

32. Hall, D.K.; Riggs, G.A.; Salomonson, V.V. Development of methods for mapping global snow cover using moderate resolution imaging spectroradiometer data. *Remote Sens. Environ.* **1995**, *54*, 127–140. [CrossRef]

33. Singh, V.P.; Singh, P.; Haritashya, U.K. (Eds.) *Encyclopedia of Snow, Ice and Glaciers*; Encyclopedia of Earth Sciences Series; Springer: Dordrecht, The Netherlands, 2011.

34. ESA—Sentinel Online. Level-2A Algorithm Overview. 2018. Available online: https://earth.esa.int/web/sentinel/technical-guides/sentinel-2-msi/level-2a/algorithm (accessed 6 December 2018).

35. Oerlemans, J.; Klok, E.J. Effect of summer snowfall on glacier mass balance. *Ann. Glaciol.* **2004**, *38*, 97–100. [CrossRef]

36. Charalampidis, C.; Fischer, A.; Kuhn, M.; Lambrecht, A.; Mayer, C.; Thomaidis, K.; Weber, M. Mass-Budget Anomalies and Geometry Signals of Three Austrian Glaciers. *Front. Earth Sci.* **2018**, *6*, 205. [CrossRef]

37. Dyurgerov, M.; Meier, M.F.; Bahr, D.B. A new index of glacier area change: A tool for glacier monitoring. *J. Glaciol.* **2009**, *55*, 710–716. [CrossRef]

38. Mattia, F.; Le Toan, T.; Souyris, J.C.; de Carolis, C.; Floury, N.; Posa, F.; Pasquariello, N.G. The effect of surface roughness on multifrequency polarimetric SAR data. *IEEE Trans. Geosci. Remote Sens.* **1997**, *35*, 954–966. [CrossRef]

39. Deroin, J.P.; Company, A.; Simonin, A. An empirical model for interpreting the relationship between backscattering and arid land surface roughness as seen with the SAR. *IEEE Trans. Geosci. Remote Sens.* **1997**, *35*, 86–92. [CrossRef]

40. Zribi, M.; Dechambre, M. A new empirical model to retrieve soil moisture and roughness from C-band radar data. *Remote Sens. Environ.* **2003**, *84*, 42–52. [CrossRef]

41. Zribi, M.; Gorrab, A.; Baghdadi, N. A new soil roughness parameter for the modelling of radar backscattering over bare soil. *Remote Sens. Environ.* **2014**, *152*, 62–73. [CrossRef]

42. Drolon, V.; Maisongrande, P.; Berthier, E.; Swinnen, E.; Huss, M. Monitoring of seasonal glacier mass balance over the European Alps using low-resolution optical satellite images. *J. Glaciol.* **2016**, *62*, 912–927. [CrossRef]

43. Floricioiu, D.; Rott, H. Seasonal and short-term variability of multifrequency, polarimetric radar backscatter of Alpine terrain from SIR-C/X-SAR and AIRSAR data. *IEEE Trans. Geosci. Remote Sens.* **2001**, *39*, 2634–2648. [CrossRef]

Sample Availability: S1 and S2 data can be downloaded from the respective webpages of Copernicus Services, Landsat data can be obtained through GLOVIS USGS. Processed S1 results are available from the authors on request.

Review

Review of Snow Data Assimilation Methods for Hydrological, Land Surface, Meteorological and Climate Models: Results from a COST HarmoSnow Survey

Jürgen Helmert [1,*], Aynur Şensoy Şorman [2], Rodolfo Alvarado Montero [3], Carlo De Michele [4], Patricia de Rosnay [5], Marie Dumont [6], David Christian Finger [7], Martin Lange [1], Ghislain Picard [8], Vera Potopová [9], Samantha Pullen [10], Dagrun Vikhamar-Schuler [11] and Ali Nadir Arslan [12]

[1]   Deutscher Wetterdienst (DWD), Offenbach 63067, Germany; Martin.Lange@dwd.de
[2]   Anadolu University, Faculty of Engineering, Department of Civil Engineering., Eskisehir 26555, Turkey; asensoy@anadolu.edu.tr
[3]   Deltares, Operational Water Management Department, Delft 2600 MH, The Netherlands; Rodolfo.AlvaradoMontero@deltares.nl
[4]   Politecnico di Milano, Department of Civil and Environmental Engineering, P.zza L. da Vinci 32, Milano 20133, Italy; carlo.demichele@polimi.it
[5]   European Centre for Medium-Range Weather Forecasts (ECMWF), Reading RG2 9AX, UK; patricia.rosnay@ecmwf.int
[6]   Météo-France—CNRS, CNRM, UMR 3589, CEN, Saint Martin d'Hères F-38400, France; marie.dumont@meteo.fr
[7]   School of Science and Engineering, Reykjavik University; Reykjavik, 101, Iceland; fingerd@gmx.net
[8]   UGA, CNRS, Institut des Géosciences de l'Environnement (IGE), UMR 5001, Grenoble 38041, France; ghislain.picard@univ-grenoble-alpes.fr
[9]   Department of Agroecology and Biometeorology, Czech University of Life Sciences Prague, Kamycka 129, Prague 165 21, Czech Republic; potop@af.czu.cz
[10] Met Office, FitzRoy Road, Exeter, Devon EX1 3PB, UK; samantha.pullen@metoffice.gov.uk
[11] Norwegian Meteorological Institute, Oslo 0313, Norway; dagrun@met.no
[12] Finnish Meteorological Institute, Helsinki FI-00560, Finland; ali.nadir.arslan@fmi.fi
*   Correspondence: juergen.helmert@dwd.de; Tel.: +49-69-8062-2704

Received: 28 September 2018; Accepted: 7 December 2018; Published: 14 December 2018

Abstract: The European Cooperation in Science and Technology (COST) Action ES1404 "HarmoSnow", entitled, "A European network for a harmonized monitoring of snow for the benefit of climate change scenarios, hydrology and numerical weather prediction" (2014-2018) aims to coordinate efforts in Europe to harmonize approaches to validation, and methodologies of snow measurement practices, instrumentation, algorithms and data assimilation (DA) techniques. One of the key objectives of the action was "Advance the application of snow DA in numerical weather prediction (NWP) and hydrological models and show its benefit for weather and hydrological forecasting as well as other applications." This paper reviews approaches used for assimilation of snow measurements such as remotely sensed and in situ observations into hydrological, land surface, meteorological and climate models based on a COST HarmoSnow survey exploring the common practices on the use of snow observation data in different modeling environments. The aim is to assess the current situation and understand the diversity of usage of snow observations in DA, forcing, monitoring, validation, or verification within NWP, hydrology, snow and climate models. Based on the responses from the community to the questionnaire and on literature review the status and requirements for the future evolution of conventional snow observations from national networks and satellite products, for data assimilation and model validation are derived and suggestions are formulated towards standardized and improved usage of snow observation data in snow DA. Results of the conducted survey showed that there is a fit between the snow macro-physical variables required for snow DA and those provided by the measurement networks, instruments, and techniques. Data availability and resources to integrate the data in the model environment are identified as the current barriers and limitations for the use of new or upcoming snow data sources. Broadening resources to integrate enhanced snow data would promote the future plans to make use of them in all model environments.

Keywords: COST Action ES1404; HarmoSnow; snow measurements; snow models; data assimilation; remote sensing

---

1. Introduction

As a major part of the cryosphere, snow is an important component of the hydrological cycle and with its unique physical properties, it is an essential environmental variable directly affecting the Earth's energy balance. Proper description and assimilation of snow information into hydrological, land surface, meteorological and climate models is therefore important to address the impact of snow on various phenomena such as hydrological monitoring, avalanche forecast, and weather forecast, to predict snow water resources and to warn about snow-related natural hazards [1–9].

Understanding the microstructural, macrophysical, thermal and optical properties of snowpack is essential [10] and there is a great need for accurate snow data at different spatial and temporal resolutions to address the challenges of changing snow conditions.

Distinctive snow properties are currently determined by traditional ground-based measurements as well as remote sensing, over a range of scales, following considerable developments in instrument technology over recent years. Snow measurements are becoming increasingly important for freshwater management, mitigation of climate changes, adaptation to new climate conditions, and risk assessments such as avalanches, floods [11], and droughts [12].

At the present time in situ measurements of the snowpack state are performed on the ground at numerous stations (e.g., WMO synoptic stations) and during intensive field campaigns (e.g., [13–17]). Simultaneous measurements of snow properties and soil properties are of further advantage [18] but only available at selected stations. However, depending on the region, in situ measurements could have a relatively coarse spatial coverage and are only representative of a limited area due to spatial heterogeneities of snow [19–21]. An increase in the number of snow measurements from national high-resolution weather networks integrated into the WMO GTS (Global Telecommunication System) and WIS (WMO Information System) would thus provide a clear benefit [22]. With the implementation of the Global Cryosphere Watch (GCW) in 2011, the WMO established a program that considers the growing demand for authoritative information on past, present and future state of the world's snow and ice resources [23]. Although GCW is global in scope, the program needs activities at all scales, including regional, national and local levels [24] and recognizes the requirements for assimilation, model development and validation.

Space-born remote sensing data have the potential to provide estimates of certain snow properties [25]. In the visible (VIS) and near infrared (NIR) spectral range space-borne remote sensors (e.g., MODIS, AVHRR, Sentinel-2) can determine the snow cover extent (SCE) and snow cover fraction (SCF) at a high spatial resolution and long time-series of these data exist (e.g., [26–33]). The observation of snow cover area is of particular value in headwaters of mountainous regions [34–36] and one can expect to obtain volume information thanks to recent advances in photogrammetry and in the availability of stereo image [37]. In addition, remotely sensed daily SCE has been shown to improve performance of hydrological models applied to various catchments in Austria [38,39], Italy [40,41], Switzerland [8,42,43], Turkey [34,44–46], Iceland [47] and the USA [48].

Optically derived snow cover products are considerably limited by the presence of clouds [49], which results in spatiotemporal gaps [25]. This limitation was quantified by [38], who found that, on average, clouds obscured 63 % of Austria in MODIS daily snow maps, from February 2000 to December 2005. Similarly, [50] found that, on average, clouds obscured 55% (MODIS Aqua) and 50% (MODIS Terra) of Po river basin (Northern Italy), for the period 2003-2012. Interestingly, they have pointed out that in the Alpine region of the basin (>1000 m.a.s.l.), the presence of clouds increases during the melting season when SCE and SCF products are most relevant: on average the percentage of cloud obstruction is 70%. Thus, cloud removal procedures are necessary to mask clouds for the snow product to be used or assimilated in hydrologic and land surface models. In the literature, different cloud removal procedures were developed based on temporal and spatial filters, see, e.g., for MODIS products [39,50–55]. In addition, digital imagery for snow extent monitoring [56] are conducted with a newly developed system for acquisition, processing and visualization of image time series from multiple camera networks [57]. These systems could connect in situ measurements and remote-sensing products and could provide SCE information in overcast conditions.

Passive microwave sensors can measure snow mass (snow water equivalent, SWE) and are not affected by illumination (night, clouds), which limits optical data during much of the high latitude snow season [58–60]. However, the spatial resolution of passive microwave SWE data is currently too coarse for many watershed-scale hydrological applications in mountainous regions [30,61], and point gauge snow data have sparse and uneven spatial coverage [62] and their accuracy is sensitive to the assumptions used, the topography, and properties of the snow pack (e.g., [63–71]). Alternatively, active microwave sensors have the potential to determine snow depth or mass from space with higher resolution but require spaceborne measurements at appropriate frequencies (Ku-band) [25] and the time resolution is more limited than for passive microwave sensors. In addition, the signal is sensitive to the snow layer properties, which complicates direct estimation of SWE from the satellite signal. For example, based on the Synthetic Aperture Radar (SAR) Interferometry technique and Sentinel-1 data, Snow Water Equivalent (SWE) temporal variations with sub-centimeter measurement accuracy can be retrieved with up to 20 m spatial resolution in any weather and sun illumination condition [72].

In addition to the determination of snow characteristics by in situ and remotely sensed measurements, another approach to obtain snow properties is to use physical or conceptual snow evolution modeling. Three major classes of snowpack models are employed for various applications [73]: single-layer snow scheme (e.g., [74]), scheme of intermediate complexity (e.g., [75]) and detailed snowpack models, which differ in the description and the parameterization of the properties inside the snowpack and the related processes [76–78]. Single-layer representations of snow thermodynamics are still used in operational NWP models [20]. In more advanced land surface schemes employed by operational models multi-layer snow options with fixed or variable numbers of layers are available [20], e.g., HTESSEL at ECMWF [79], JULES at the Met Office [80], ISBA-ES in SURFEX [81], and TERRA in the ICON model at DWD [82]. Detailed snowpack models include in addition state variables for snow microstructure, namely grain sizes and shapes in layers [20,83]. However, continuous estimates of the snow state from numerical model predictions are still limited by scarcity and uncertainties in meteorological forcing data (see [84] for an example) and model structural problems for snow processes in land surface models [25,85–87].

Assimilation of remotely sensed snow-related observations and ground-based snow measurements has been proven to be an effective method to improve hydrological and snow model simulations [88–93]. Therefore, the potential of data assimilation methods to consistently improve model simulations of the snow state by assimilation of measurements from in situ as well as from remote sensing has received continuously increasing attention [25,59,88–102]. With data assimilation (DA hereinafter) techniques, an improvement of the simulated snow properties from numerical models can be obtained by the combination of observational datasets with numerical model predictions and consideration of the uncertainties of observed and modeled variables [103]. Several studies report the assimilation of in situ snow observations [62,101,103–105] even for operational applications [2,106,107]. A number of snow DA experiments taking into account remotely sensed snow properties have been performed with different observational datasets, including snow covered area and SWE [59,88,90,91,100,108,109].

However, visible or near-infrared observations do not allow assimilation updates under cloudy conditions and the updates depend on the snow depletion curve used to relate SCE or SCF to SWE [25]. Direct assimilation of SWE from passive microwave remote sensing data exists [88,100,110] but radiance assimilation may be more effective [25,111–118]. The latter approach is indeed able to overcome the difficulty arising from the non-unique and complex relationship linking the passive microwave signal and several snow properties (e.g., density, grain size/microstructure parameters, temperature and wetness). To assimilate radiance, radiative transfer modeling is needed to play the role of the 'observation operator' in the DA scheme, that is to relate the snow properties predicted by the dynamical snow scheme to the remotely-sensed observed variables as well as to provide the associated uncertainties [119]. Many such numerical models have been developed and evaluated over the last decade for the passive microwave such as HUT [120], MEMLS [121], DMRT-QMS [122], DMRT-ML [123]. Although their performances appear to be comparable (e.g., [124]), the formulations and parameterizations used in each model are diverse. This apparent paradox is only partially understood (e.g., [125,126]), which has motivated the development of a uniform, harmonized, modeling platform called the Snow Microwave Radiative Transfer Model (SMRT, [127]). This new model is also able to better represent the snow microstructure, which currently remains the main bottleneck for SWE estimation [114,128].

The European Cooperation in Science and Technology (COST) promoted and funded the Action ES1404 called "A European network for a harmonized monitoring of snow for the benefit of climate change scenarios, hydrology, and numerical weather prediction," or "HarmoSnow" for short. The HarmoSnow project (2014-2018) coordinates efforts towards harmonized snow data processing and handling practices by promoting new observing strategies, bringing together different communities, facilitating data transfer, upgrading and enlarging knowledge through networking, exchange and training, and linking them to activities in international agencies and global networks [129]. Due to the large heterogeneity of methods and tools for manual measurement of snow and their assimilation in numerical models one of the first activities of HarmoSnow was to carry out surveys to obtain an updated picture of the existing variety of a) snow measurement practices and instrumentations, and b) the data assimilation methods and snow data processing used in NWP, hydrology, and climate studies by the European institutions. The results of the first survey are published in [11]. This paper aims to assess the current situation and understand the diversity of usage of snow observations in DA, forcing, monitoring, validation, or verification within NWP, hydrology, snow and climate models. Our findings are based on the responses from the community to the second survey, on snow DA methods and processing, and on literature review.

2. European Survey on Usage of Snow Observations in Data Assimilation, Forcing, Monitoring, Validation, or Verification

The survey was conducted via an online questionnaire from September 2015 to December 2017 on the COST HarmoSnow website. This questionnaire (see Supplementary Material: COST ESSEM 1404 working group 3 survey: Questionnaire and results) was compiled by COST HarmoSnow experts in snow modeling and data assimilation and distributed across the COST, EUMETSAT H-SAF and GCW member networks. The questionnaire was answered by 51 participants from 31 countries. The survey consists of 32 questions in six sections and one text box for additional comments (see Supplementary Material: COST ESSEM 1404 working group 3 survey: Questionnaire and results), which are also available at the COST HarmoSnow website. Most questions used multiple choice answers. This procedure ensures clear answers and space for further explanations was provided. A weighting of the answers was not made, but we are aware of their heterogeneity in terms of institutional representativeness and implications for the representativeness of our derived conclusions. The evaluation was performed manually.

3. Results

The results of the survey are presented in this section, grouped into sub-sections according to the thematic topic that the survey questions address. Description and interpretation of the results is therefore separated into: the range of participants in the survey; the modeling environments used by participants, the data assimilation methods used; snow observations used in these DA methods; treatment of background and observation errors; quality control methods used; data exchange policy and requirements; and plans for future observation use.

3.1. Participating Countries and Institutions

The distribution of the number of answers among the countries is shown in Figure 1. It shows that all responses were from countries from the northern hemisphere, with the majority from central Europe (27). With the Nordic countries, Russian Federation, the USA and Canada, most countries in the boreal forest belt answered the survey. These countries contain regions that always have seasonal snow in northern hemisphere winter, while for countries in central and southern Europe the number of days with snow cover is more variable and depends on several factors. For reporting countries having relatively lower latitude and high altitudes, snow in the climatological mean is limited to the mountains, however it is an important factor for meteorological and hydrological applications. Most European countries involved in the COST action provided at least one answer, thus the dataset of answers provides a solid base for analysis to obtain an overview on the utilization of snow data in NWP systems, hydrological models, special snow models and treatment of snow in other model environments (Figure 2).

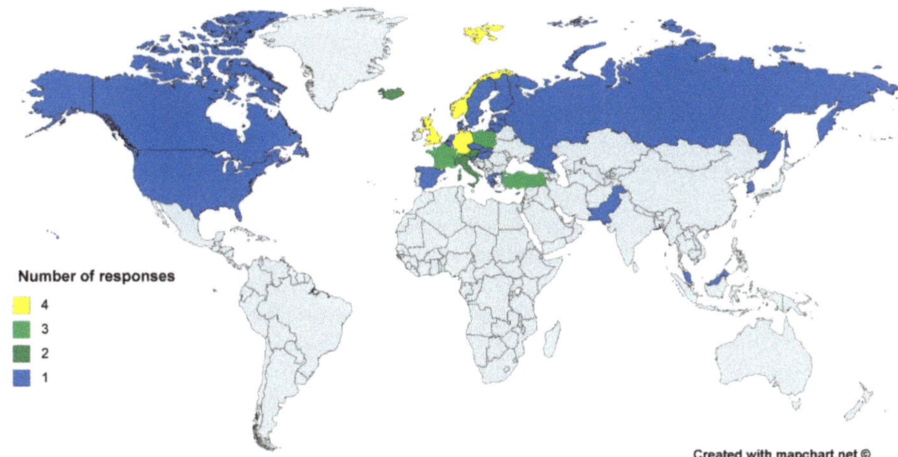

Figure 1. Geographical distribution of number of responses in the survey.

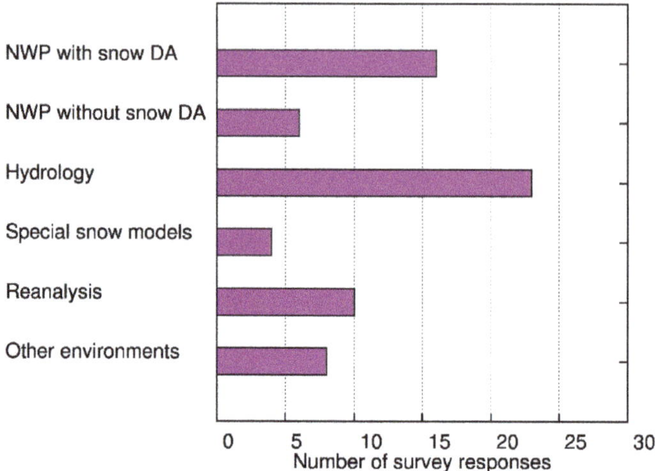

Figure 2. Distribution of modeling environment within the questionnaire.

In general, multiple responses from each country were not expected since most snow DA activity takes place within the national met service of that country, which is a fairly standard situation in most European countries. The national weather services in the countries with huge territories (Russia, Canada and the USA) include regional institutions with their own capabilities to produce local weather forecasts.

3.2. Modeling Environment, Model Domain and Resolution

The assessment results have been partitioned according to the type of modeling environment that the respondent has identified as using. Among respondents, 16 institutes use numerical weather prediction models (full or limited area) with DA, six without DA, 23 institutes use hydrological models (e.g., conceptual, operational, snowmelt models, runoff models, etc.), 10 institutes use reanalysis and four institutes employ special snow models. In eight institutes other (miscellaneous) models (e.g., snow cover, land surface, multi-layer snow), with snow observations are used (Figure 2). In addition to meteorological and hydrological services, 11 universities and two companies participated in the survey. The resolutions of the models span from the global scales down to kilometer scale resolutions and even to the river catchment areas according to the modeling environment, clearly proving the declared importance of snow observations over a range of spatial scales.

3.3. Data Assimilation Methods

There are differences in the snow DA methods used by the various model environments as well as in the update frequency of snow observations and the required time interval for consideration of the measurement. Depending on their degree of complexity, DA techniques are characterized by different performances. The sequential DA techniques are widely used for real-time applications. They sequentially update the model state using observational data as they become available [102]. Basic approaches are based on direct insertion (DI) methods [104,108,130] or Cressman interpolation [2,131,132]. Other approaches include optimal interpolation (OI) schemes [103,107,133,134] and the nudging method [135,136], which take into account the observational uncertainty [102].

The Kalman filters [137] are approaches based on least-squares analysis method. The standard version of the Kalman Filter (KF) [138] still depends on the assumption of system linearity since it explicitly takes into account the dynamical nature of model and observation errors, which evolve with time, to produce a statistically optimal model state estimates for linear systems. The Extended Kalman Filter (EKF) [139] allows consideration of nonlinear dynamic models using a linearized statistical approach [59,110]. With the Ensemble Kalman Filter (EnKF) the inaccuracy of the linearization procedure, which affects the filter performance due to possible strong model nonlinearities [140], can be avoided [141]. An ensemble of possible model realizations is needed based on the Monte Carlo approach [142] to determine the error estimates instead of a model linearization. Other, more sophisticated methods, that include for example particle filter [143], are also used for snow data assimilation in hydrology. Similar to the EnKF, the particle filter (PF) is a sequential Monte Carlo simulation [43,90,97,117,143–147] and accounts for uncertainties in the forcing data, model structure and observations. However, in contrast to the EnKF the PF does not depend on the assumptions of Gaussian distribution of errors or the assumption of Gaussian joint probability density function (PDF) on the state variables and observations [19,148]. This allows the PF to characterize the full probability distribution of state variables and consequently their uncertainties more accurately by resampling sets of state variables, i.e., particles with higher posterior weights, as opposed to the linear model state updating of the EnKF [19]. Another DA approach is the variational Moving Horizon Estimation (MHE), which optimizes an objective function within an assimilation window using numerical approximations [149]. This type of methods is also applied in Model Prediction Control applications when the assimilation window is shifted to the predictive horizon [150]. Recently, [151] extended this approach to consider multi-parametric conditions in consideration with snow DA.

Despite the previous drawbacks, EnKF techniques have been widely implemented to process snow observation data [19,20,62,88,111,152–156]. Also in snow hydrology, an increasing number of studies confirm the advantages of EnKF as a data assimilation method, which improves the accuracy of hydrological simulations through the assimilation of snow-related observations [25,62,88,89,101,102,111,152,157–160]. Other studies have shown the benefits of assimilation of snow using EKF [59,110] or PF [19], while [100] applied the MHE to evaluate different satellite products, including snow observations, within the DA procedure.

The results of the survey show that the OI method and EnKF are the most commonly used methods (Figure 3). The survey revealed that snow DA for NWP mostly relies on optimal interpolation (OI) schemes [107] or Cressman interpolation [131], on the other hand Kalman filters [137] or EnKF methods [141] are generally used for hydrological applications. Other answers in the survey include Moving Horizon Estimation, Nudging, Asynchronous EnKF, Bias Detecting Ensemble, simple exchange of initial values, and simple update method. However, it should be noted that for complex multi-layered snow models, the application of conceptually simple DA schemes is not straightforward due to possible model spin-up behavior resulting from physical inconsistencies among state variables [97,161].

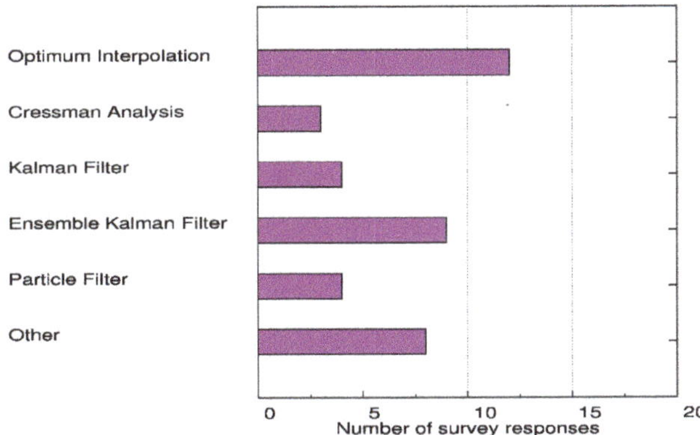

Figure 3. Data assimilation methods.

The survey results confirm the existing gap of applied sophistication in DA methods for operational NWP, i.e., the DA methods used for snow analyses in NWP are much simpler than the state of the art in DA [20] and lag behind the level of sophistication used for the initialization of other surface variables (e.g., soil moisture). Furthermore, operational NWP systems assimilate snow depth from in situ ground measurements and satellite-derived snow extent [2,134,162] but SWE is not considered during the assimilation cycle [20]. Recent effort to implement advanced techniques such as the particle filter [163], show promising results, although the path toward operational use is long [97].

3.4. Snow Observations in Data Assimilation through Different Models

Snow observations from SYNOP and additionally ground-based measurements are the most important data sources for NWP and hydrologic models (Figure 4). For the latter, ground-based remote sensing data are also very important. The most important snow parameters used in DA are snow depth and SWE, which are processed by incremental update for NWP or update of absolute values in hydrologic and other snow models.

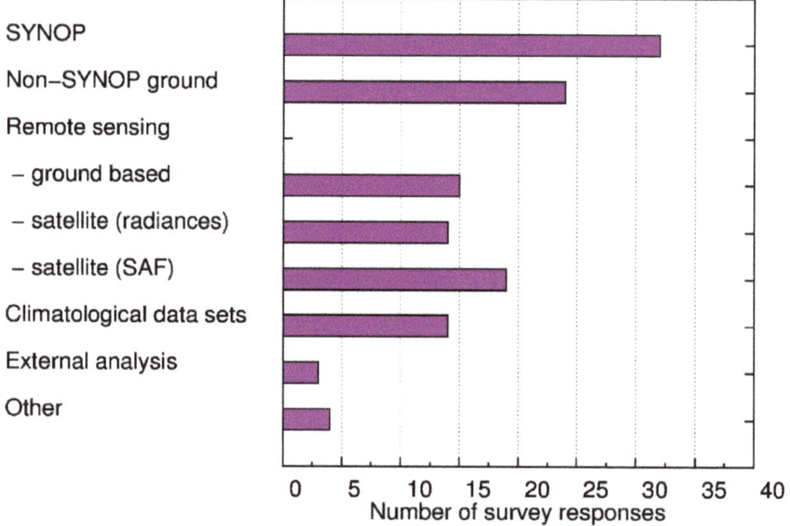

Figure 4. Snow observations and products used in the modeling system.

According to the answers, snow height (depth) is the most preferable information for DA in NWP models. For these data, the importance of an active reporting of snow-free conditions (zero snow depth) in the SYNOP messages together with the exchange of non-GTS stations data is crucial. For hydrologic models, both snow height (depth) and SWE are popular for use in DA, as complementary products to streamflow assimilation. For this group of models, forcing variables (precipitation, temperature) can also be used in the assimilation process to update state variables. The main snow variable analyzed in almost every type of model is SWE.

Most hydrologic model users that responded to the survey use ground-based remote sensing measurements, while this is not the case for NWP or reanalysis users. The ground-based measurement systems include ultrasonic or laser distance sensors, photogrammetry, COSMIC neutron sensors and others. Further details on in situ snow measurements are given in the results of the parallel COST HarmoSnow survey on in situ snow measurement practices and techniques, [11].

Preprocessed remote sensing satellite products are also often used in both NWP and hydrology. Satellite radiances are used much less and climatological data are appropriated for hydrological applications. Additional data, used by survey participants include external snow analysis or multi-sensor satellite products. Preprocessed snow products are used in all model environments but these products have special importance in NWP without DA, reanalysis and miscellaneous models. The used products are, e.g., from IMS snow cover, satellite (MODIS, SEVIRI, AVHRR), SAF (H-SAF, LSA-SAF), NWP-based snow analysis or reanalysis.

The process of analyzing variables is mostly incremental update of first guess from model forecast for NWP models and update of absolute values for hydrological models. However, the modelers use both processes together in some applications. Model forecasts are the main background field used in snow data assimilation for all model types. However, a very limited number of answers include pre-analysis or external analysis and climatology as a background field.

Independently of the modeling environment most DA systems perform a snow analysis every 24 h (Figure 5). This is important for the assimilation of remote sensing data since not all satellite products are available on a daily basis. Even daily products for SWE based on passive microwave remote sensing data have a coarse resolution (in the order of tens of kilometers) while SWE products from active microwave sensors can reach a resolution of tens of meters but are only available every few days (e.g., 35 days with ERS and Environmental Satellite (ENVISAT) in the past decades and nowadays six days with Sentinel 1).

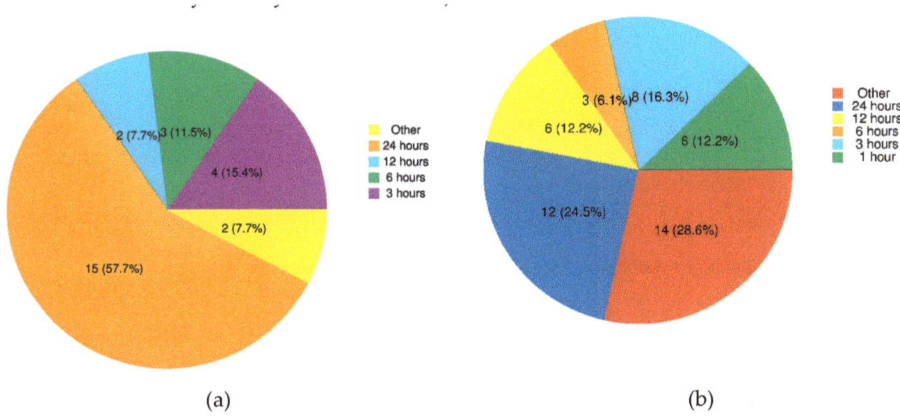

Figure 5. Update frequency for snow data assimilation (a) and observation latency (b).

The observation data latency, i.e., the time from the measurement acquisition to the availability in a numerical model is another important parameter, which has to be considered for time-critical applications. More than 50% of the survey answers indicate that the observation latency should not exceed 24 h. For research applications or in climate studies, a longer latency might be acceptable. However, more than 25% of the answers show that the observations should be available within 3 h, which is a strong constraint for the observation data processing and exchange.

3.5. Background and Observation Error Estimations Used in Snow Data Assimilation

The background error estimates are done either by distance weighting or taken as a fixed value in most of the NWP models, the former is more commonly used compared to the latter. The variance of ensembles is another method used in limited applications. A few institutes working on NWP also indicated that background errors are not accounted for in their system. For the other model communities, the answers are more varied and include no estimate, fixed value, distance weights, stochastic noise, defined algorithms with clearly more emphasis on variance of ensembles, which is most likely due to the choice of EnKF for DA methodology.

According to the survey there is no standard approach for observation error estimates. Generally, standard deviations or fixed values according to the measurement errors (in principle different for different observation types) are used in NWP models. Some of the institutes do not use error estimates assuming uncorrelated observed data except for the anomalous observations identified and rejected by quality control procedures. In the other modeling environments, observation errors are defined by measurement errors, standard deviations, confidence intervals, rough estimates, stochastic noise, error covariance matrices, or error estimates are simply not accounted for.

Observation error specification has a large impact on DA efficiency. Figure 6 illustrates the two-meter air temperature forecast (range 12-h and 3-day) difference from December 2016 to February 2017, between a test experiment, where the snow observation error was doubled in the ECMWF snow DA, and a reference experiment, using the ECMWF operational system [134]. It shows generally colder conditions in the northern hemisphere. Doubling the observation error gives relatively more weight to the model background in the test experiment compared to the reference experiment. Since the ECMWF model tends to overestimate snow, this results in more snow on the ground in the test than in the reference experiment, and therefore generally lower air temperature forecasts. Slight and noisy differences in non-snow-covered areas are non-significant and due to the fact that the test and the control experiments differ.

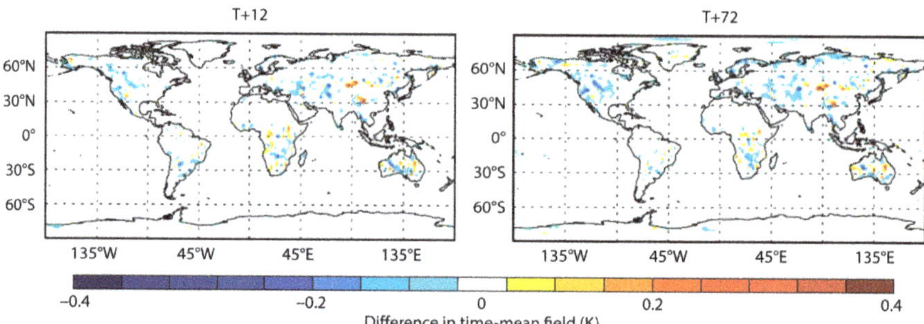

Figure 6. Impact, shown as mean temperature difference in K (01 December 2016 to 28 February 2017), of doubling snow observation error in the ECMWF snow data assimilation system used for NWP, on two-meter air temperature 12-h and 72-h forecasts.

3.6. Quality Control of Snow Observations or Products

One of the important features of DA systems is the quality control of the data [164]. It is performed by using previous model forecasts for comparison with observations. This allows identification and elimination of spurious data. Furthermore, it is possible to calibrate observing systems and identify biases or changes in observation system performance when this comparison is performed repeatedly [165].

Quality control of snow observations and products is performed in the large majority of the model environments used in this survey (Figure 7.). Filtering of outliers, manual and automatic treatment of missing data or implausible values is used in all model environments with different levels of sophistication. DA in NWP is used for this purpose, as some responses from the survey show.

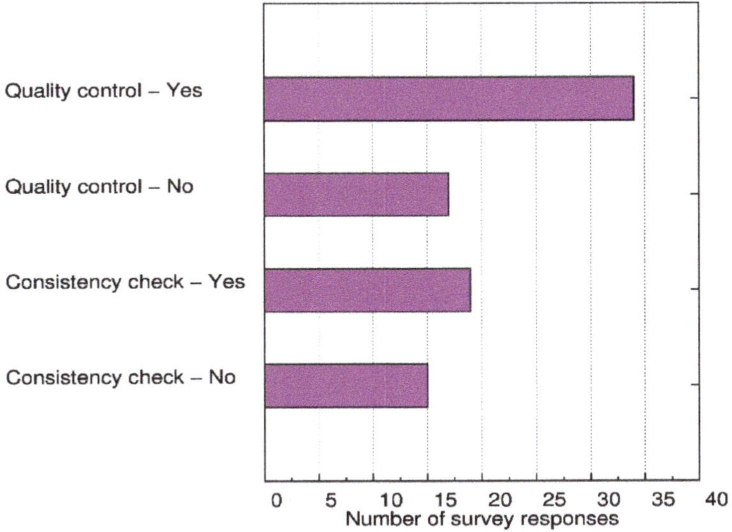

Figure 7. Quality control and consistency check of snow observations or products.

Consistency or sanity checks are used to examine whether the observation absolute value or rate of change with time is physically realistic or not [165]. In addition, buddy checks are used for the comparison of observations close together and background checks consider a realistic change of the observation in comparison to the model prediction [165]. In comparison with a quality control, the number of institutes performing a snow data consistency check is lower. For this data preprocessing manual and automatic methods exist, based on basic physical principles, where the snow cover field is of particular importance. Consistency of snow height with the existence of snow cover is one of the commonly used checks, i.e., check if snow cover is present where observations of non-zero snow depth exist.

3.7. Data Exchange Policy and Access Requirements for the Observations

The survey indicates that two thirds of all answers were positive towards the possibility of snow data exchange with other groups. The NWP community seems more flexible compared to other modeling groups in this sense. This should be moderated by the fact that a relatively higher spatial resolution and catchment scale of hydrological applications could be a constraint on the feasibility of data exchange. In most cases GTS for NWP models and FTP protocol for the other models is required for data access, but web access or central data hubs are also used.

3.8. The Plans to Use the New or Upcoming Observation Sources

Concrete plans for using new or upcoming data sources of snow observations exist for all model environments, in particular for NWP with DA, hydrology, and reanalysis. In detailed answers of the survey, the use of more satellite data (optical, microwave) and also more ground-based remote sensing data, GPS or COSMIC ray sensors, or additional non-SYNOP networks are of interest. Current barriers and limitations for the use of these data are primarily data availability and lack of resources needed to integrate the data into the model environment. Survey responses showed these barriers to be common across model environments used.

4. Summary and Discussion

According to the outcomes of the literature review, surveys and the action ES1404 itself together with HarmoSnow field campaigns and from snow HarmoSnow DA school activities, the key messages have been recognized and special attention is given to them in this section. These key items are (i) using conventional snow observations from national networks for DA and model validation, (ii) sustainable ways to create snow products for users by combining remote sensing and conventional snow observations with modeling results and (iii) snow observations errors for data assimilation and modelling systems.

4.1. How to Get and Use Conventional Snow Observations from National Networks for Data Assimilation and Model Validation

Data assimilation systems employed in model environments for numerical weather prediction, hydrology, or special snow models make extensive use of present-day measurement networks. The range of measurement instruments and techniques in use is also indicated by the results of the parallel considering COST HarmoSnow survey on in situ snow measurements [11]. The literature review reveals the importance of conventional snow observations for DA and model calibration and validation, and the potential benefits of obtaining additional observations from national networks, according to the survey responses, concrete plans for using new or upcoming data sources of snow observations exist for use in snow models of NWP, hydrology or other modeling environments.

Some strategies exist, or are underway, towards an improved and more extended usage of conventional snow observations to include observations from high-resolution national networks into NWP, hydrological and climate models, as the availability, and therefore use, of such data very limited [22,134]. They include the following: (1) The WMO "Snow Watch" initiative, which recognises the importance of near-real-time in situ observations of snow cover and snow depth to the global observing system. The Snow Watch action has secured approval by Executive Council 69 for an amendment to [166] in order to make daily reporting of snow cover and snow depth a mandatory requirement for all stations with the capability to do so. Of particular note, this explicitly includes the requirement to report values of zero snow depth, when snow is not present, in order to provide valuable additional information for assimilation into weather forecasting models. (2) The ECMWF data assimilation study with additional snow data [22,134], clearly demonstrates the benefits to NWP forecast accuracy from assimilation of additional national network snow reports. (3) The monitoring of SYNOP station snow depth reports to detect problems in the snow analysis. A continuous monitoring allows also to identify progress in observation availability and frequency of reporting, which becomes more important with a broader application of automatic snow measurements [167].

4.2. Sustainable Ways to Create Snow Products for Users by Combining Remote Sensing and Conventional Snow Observations with Modeling Results

Until now, spaceborne derived snow products are not widely applied in operational NWP systems. In situ measurements are used to retrieve snow height for data assimilation, since for SWE the satellite products still not meet the requirements of DA in NWP. For SCE a number of combined and operational products exist, which includes also in situ measurements [168] and can be used to constrain the model SCE but SWE is the most interesting variable. Therefore, a number of blended satellite products have been developed, which merge visible, near-infrared, and passive microwave observations [169–172] and could be used for DA. Microwave satellite observations are combined with conventional in situ observations in some products (H-SAF), while optical satellite observations together with conventional in situ observations are assimilated into NWP models. The ESA GlobSnow project is another example providing combined products for models. It was shown recently by [94] that the coarse resolution of space-borne radiometers (in the order of tens of kilometers) for existing SWE products (H-SAF and GlobSnow) can be improved by assimilation together with in situ observations of snow depth, where the improved resolution enhances spatial details in the retrieved SWE. For hydrology, the DA is also very important for the implementation of spaceborne snow products used in streamflow forecasts. A variational method based on Moving Horizon Estimation (MHE) is used in [16], in application to the conceptual rainfall-runoff model HBV. Snow cover extent (SCE), snow water equivalent (SWE), soil moisture (SM) and in situ measurements of streamflow data were assimilated using large assimilation windows of up to one year. For the first time, H-SAF products were used for hydrological forecasting systems and their added value was verified. Although blended satellite products could serve in filling gaps of observation data or providing validation data, ideally, blending of products should happen only within the DA process due to the preferred separation of observation sources. Furthermore, blended products often contain information from models that have different assumptions than the model, which uses these products in assimilation.

4.3. Snow Observations Errors for Data Assimilation and Modelling Systems

Snow observations and products are subject to quality control as well as consistency checks, which are performed manually or automatically in the large majority of the model environments used in this survey. Furthermore, the observation error of snow measurements consisting of instrument error and representativeness error (e.g., [173,174] is an important parameter used in snow DA. Results from the survey show that if an estimation of the observation error of snow measurements for DA is used, a prescribed constant value is chosen in many cases. The observation error sensitivity study performed with the ECMWF snow DA system showed the impact of this parameter on the global NWP forecast skills (Section 3.5, Figure 6). Since the observation error consists of two components for the derivation of this error from reported values in literature, the context of the measurements has to be taken into account [159,175–178].

The detailed review and assessment of the survey carried out by measurement communities [11] report on the quality of data and potential problems and provide valuable feedback for modelers on instrument errors. For instance, the results from the COST HarmoSnow field campaign in Iceland [16] showed that for SWE the observation error as relative standard deviation of 10% is possible if a suitable amount of measurements (i.e., minimum 3) is performed. However, this only account for the instrument error. In complex and windy terrains the total observation error can be much larger, which reflects the limited representativeness of snow measurements.

5. Conclusions

Based on the practices of a number of countries, review of the literature, and evaluation of the survey on use of snow observations in the modelling environment, conclusions can be drawn on: (i) the status and future evolution of conventional snow observations from national networks and satellite products for DA and model validation, including availability, error characteristics and reliability, towards an improved usage of conventional snow observations from national networks for data assimilation and model validation; (ii) the review of the methods to combine remote sensing and conventional snow observations with modeling results for user applications; and (iii) snow observations errors for data assimilation and modelling systems.

The results of the survey show that the measurement networks, instruments, and techniques are exploited well by existing DA systems and used in model environments for NWP, hydrology, or special snow models. The survey reveals that there is a fit between the snow macro-physical variables required for DA and those provided by the measurement environment, since snow depth, snow presence, snow density and SWE are the most measured variables. It is also important to take into consideration that in many cases these variables are measured with different instruments and techniques, in particular snow depth and SWE. On the other hand, developments in DA systems are necessary to exploit the evolving capabilities of the observing systems, and vice versa. The increasing automation of the measurements requires enhanced data management in the DA system (quality control, consistency). There is a requirement for remotely sensed snow depth or SWE observations from satellite, to provide snow data in regions with sparse measurement networks, but this necessitates developments in instrument technology (e.g., automatic measurement of snow microstructural properties) and also development of DA systems in order to make use of such observations. There are also concrete future plans on using enhanced snow observations for all model environments, in particular for NWP with DA, hydrology, and reanalysis. Data availability and resources to integrate the data in the model environment are the current barriers and limitations for the use of new or upcoming snow data sources independent of the model environment used.

The further outcomes from the survey support new, innovative and upgraded observing strategies; enhanced usage of snow data for scientific research and applications; a broader overview and easier access to existing snow measurements and snow model data for the benefit of different applications, such as NWP models, hydrological, climatology and climate change research. Further support of these aims is provided by the related COST HarmoSnow activities of the parallel survey on snow measurements and the training school on data assimilation in Europe [179]. The monitoring of floods, droughts, snow avalanches and hydropower production could benefit from improved real-time snow measurements for assimilation into operational prediction models to improve hydrological, meteorological and climate forecasting while a further integration and harmonization of the European snow network into global networks (e.g., WMO GCW) supports the strengthening WMO and EUMETSAT activities on snow observations. The main scientific impact will emerge from improved snow and weather products via better knowledge of snow properties and their evolution. It will induce a lasting structural improvement of the interaction between participating communities, thus very relevant for the Intergovernmental Panel on Climate Change (IPCC) and Copernicus (Global Monitoring for Environment and Security). Policy and decision makers at all levels from local safety to global environment policy will benefit from improved knowledge on current and future snow cover and climate conditions.

Supplementary Materials: COST ESSEM 1404 working group 3 survey: Questionnaire and results are available at http://www.harmosnow.eu/index.php?page=WG3.

Author Contributions: Conceptualization, J.H., A.S.S., P.d.R., S.P., D.C.F, C.D.M., R.A.M., M.L., M.D., G.P., V.P., D.V.-S, A.N.A.; Methodology, J.H., A.S.S., P.d.R., S.P., D.C.F, C.D.M., R.A.M., M.L., M.D., G.P., V.P., D.V.-S, A.N.A.; Writing (original draft preparation), A.S.S., J.H., P.d.R., S.P., D.C.F, C.D.M., R.A.M., M.L., M.D., G.P., V.P., D.V.-S, A.N.A.; Writing (review and editing), J.H., A.S.S., P.d.R., S.P., D.C.F, C.D.M., R.A.M., M.L., M.D., G.P., V.P., D.V.-S, A.N.A.

Funding: This research is part of the COST Action ES1404 activity funded by the COST programme.

Acknowledgments: We thank all of the participants of the COST Action ES1404 for their support and fruitful comments to the results. We also thank all of the respondents to the survey who provided the material for this study. We are grateful to Sylvain Joffre, who had a key role in planning and initiating the COST Action ES1404.

Conflicts of Interest: The authors declare no conflict of interest.

References

1. Sui, J.; Koehler, G. Rain-on-snow induced flood events in southern Germany. *J. Hydrol.* **2001**, *252*, 205–220. [CrossRef]
2. Drusch, M.; Vasiljevic, D.; Viterbo, P. ECMWF s global snow analysis: Assessment and revision based on satellite observations. *J. Appl. Meteorol.* **2004**, *43*, 1282–1294. [CrossRef]
3. Finger, D.; Heinrich, G.; Bauder, A. Projections of future water resources and their uncertainty in a glacierized catchment in the Swiss Alps and the subsequent effects on hydropower production during the 21st century. *Water Resour. Res.* **2012**, *48*, W02521. [CrossRef]
4. Viviroli, D.; Archer, D.R.; Buytaert, W.; Fowler, H.J.; Greenwood, G.B.; Hamlet, A.F.; Huang, Y.; Koboltschnig, G.; Litaor, M.I.; Lopez-Moreno, J.I.; et al. Climate change and mountain water resources: Overview and recommendations for research, management and policy. *Hydrol. Earth Syst. Sci.* **2011**, *15*, 471–504. [CrossRef]
5. Freudiger, D.; Kohn, I.; Stahl, K.; Weiler, M. Large-scale analysis of changing frequencies of rain-on-snow events with flood-generation potential. *Hydrol. Earth Syst. Sci.* **2014**, *18*, 2695–2709. [CrossRef]

6.   Fayad, A.; Gascoin, S.; Faour, G.; Lopez-Moreno, J.I.; Drapeau, L.; Le Page, M.; Escadafal, R. Snow Hydrology in Mediterranean Mountain Regions: A. Review. *J. Hydrol.* **2017**, *551*, 374–396. [CrossRef]

7.   Lafaysse, M.; Cluzet, B.; Dumont, M.; Lejeune, Y.; Vionnet, V.; Morin, S. A multi physical ensemble system of numerical snow modeling. *Cryosphere* **2017**, *11*, 1173–1198. [CrossRef]

8.   Etter, S.; Addor, N.; Huss, M.; Finger, D. Climate change impacts on future snow, ice and rain runoff in a Swiss mountain catchment using multi-dataset calibration. *J. Hydrol. Reg. Stud.* **2017**, *13*, 222–239. [CrossRef]

9.   Sturm, M.; Goldstein, M.A.; Parr, C. Water and life from snow a trillion dollar science question. *Water Resour. Res.* **2017**, *53*, 3534–3544. [CrossRef]

10.   Singh, P.; Singh, V.P. *Snow and Glacier Hydrology*; Kluwer Academic Publishers: Dordrecht, The Netherlands, 2001; p. 104.

11.   Pirazzini, R.; Leppänen, L.; Picard, G.; Lopez-Moreno, J.I.; Marty, C.; Macelloni, G.; Kontu, A.; von Lerber, A.; Tanis, C.M.; Schneebeli, M.; et al. European In-Situ Snow Measurements: Practices and Purposes. *Sensors* **2018**, *18*, 2016. [CrossRef]

12.   Hatchett, B.J.; McEvoy, D.J. Exploring the Origins of Snow Drought in the Northern Sierra Nevada. California. *Earth Interact.* **2018**, *22*, 1–13. [CrossRef]

13.   Cline, D.; Yueh, S.; Chapman, B.; Stankov, B.; Gasiewski, A.; Masters, D.; Elder, K.; Kelly, R.; Painter, T.H.; Miller, S.; et al. NASA Cold Land Processes Experiment (CLPX 2002/03): Airborne Remote Sensing. *J. Hyrdometerol.* **2009**, *10*, 338–346. [CrossRef]

14.   COST ESSEM 1404, Field Campaign in Chopok, Slovakia, 15–16 February 2016. Available online: http://www.harmosnow.eu/dissemination/reports/Field%20campaign%20SK.pdf (accessed on 13 December 2018).

15.   COST ESSEM 1404, Field Campaign in Erzurum, Turkey, 1–3 March 2016. Available online: http://www.harmosnow.eu/dissemination/reports/Field_Campaign_Erzurum_2016.pdf (accessed on 13 December 2018).

16.   COST ESSEM 1404, Field Campaign in Reykjavik, Iceland, 28 February 2 March 2017. Available online: http://www.harmosnow.eu/dissemination/reports/COST_2nd_field_campaign_report.pdf (accessed on 13 December 2018).

17.   Kim, E.; Gatebe, C.; Hall, D.; Newlin, J.; Misakonis, A.; Elder, K.; Marshall, H.; Hiemstra, C.; Brucker, L.; De Marco, E.; et al. NASA's snowex campaign: Observing seasonal snow in a forested environment. *J. Geophys. Res. Atmos.* **2017**, 1388–1390. [CrossRef]

18.   Potopová, V.; Boroneat, C.; Možný, M.; Soukup, J. Driving role of snow cover on soil moisture and drought developing during the growing season in the Czech Republic. *Int J. Climatol.* **2016**, *36*, 3741–3758. [CrossRef]

19.   Dechant, C.M.; Moradkhani, H. Improving the characterization of initial condition for ensemble streamflow prediction using data assimilation. *Hydrol. Earth Syst. Sci.* **2011**, *15*, 3399–3410. [CrossRef]

20.   Essery, R. Snowpack Modeling and Data Assimilation. ECMWF-WWRP/THORPEX Workshop on Polar Prediction. Available online: https://www.ecmwf.int/sites/default/files/elibrary/2013/13948-snowpack-modelling-and-data-assimilation.pdf (accessed on 13 December 2018).

21.   Dong, C. Remote sensing, hydrological modeling and in-situ observations in snow cover research: A review. *J. Hydrol.* **2018**. [CrossRef]

22.   De Rosnay, P.; Mallas, I.; Gospodinov, I. Additional snow depth reports from Bulgaria: Data assimilation and recommendations. *ECMWF Res. Memorandum* **2016**, RD16-178.

23.   Key, J.; Goodison, B.; Schöne, W.; Godøy, Ø.; Ondráš, M.; Snorrason, Á. A Global Cryosphere Watch. *ARCTIC* **2015**, *68*, 48–58. [CrossRef]

24.   World Meteorological Organization. *Global Cryosphere Watch (2012) First Implementation Meeting*; Final Report; World Meteorological Organization: Geneva, Switzerland, 2011.

25.   De Lannoy, G.J.M.; Reichle, R.H.; Arsenault, K.R.; Houser, P.R.; Kumar, S.; Verhoest, N.E.C.; Pauwels, V.R.N. Multiscale assimilation of Advanced Microwave Scanning Radiometer–EOS snow water equivalent and Moderate Resolution Imaging Spectroradiometer snow cover fraction observations in northern Colorado. *Water Resour. Res.* **2012**, *48*, W01522. [CrossRef]

26.   Hall, D.K.; Riggs, G.A.; Salomonson, V.V. Development of Methods for Mapping Global Snow Cover Using Moderate Resolution Imaging Spectroradiometer Data. *Remote Sens. Environ.* **1995**, *54*, 127–140. [CrossRef]

27.   Klein, A.G.; Hall, D.K.; Riggs, G.A. Improving snow cover mapping in forests through the use of a canopy reflectance model. *Hydrol. Process.* **1998**, *12*, 1723–1744. [CrossRef]

28. Metsämäki, S.; Vepsäläinen, J.; Pulliainen, J.; Sucksdorff, Y. Improved linear interpolation method for the estimation of snow-covered area from optical data. *Remote Sens. Environ.* **2002**, *82*, 64–78. [CrossRef]
29. Akyurek, Z.; Surer, S.; Beser, Ö. Investigation of the snow-cover dynamics in the Upper Euphrates Basin of Turkey using remotely sensed snow-cover products and hydrometeorological data. *Hydrol. Process.* **2011**, *25*, 3637–3648. [CrossRef]
30. Dietz, A.J.; Wohner, C.; Kuenzer, C. European Snow Cover Characteristics between 2000 and 2011 Derived from Improved MODIS Daily Snow Cover Products. *Remote Sens.* **2012**, *4*, 2432–2454. [CrossRef]
31. Parajka, J.; Bezak, N.; Burkhart, J.; Hauksson, B.; Holko, L.; Hundecha, Y.; Jenicek, M.; Krajčí, P.; Mangini, W.; Molnar, P.; et al. MODIS snowline elevation changes during snowmelt runoff events in Europe. *J. Hydrol. Hydromech.* **2018**, *67*, 101–109. [CrossRef]
32. Şorman, A.A.; Uysal, G.; Şensoy, A. Probabilistic snow cover and ensemble streamflow estimations in the Upper Euphrates Basin. *J. Hydrol. Hydromech.* **2018**, *67*, 82–92.
33. Simon, G.; Grizonnet, M.; Klempka, T.; Salgues, G. Algorithm theoretical basis documentation for an operational snow cover product from Sentinel-2 and Landsat-8 data (Let-it-snow). *Zenodo* **2018**. [CrossRef]
34. Tekeli, A.E.; Akyurek, Z.; Sorman, A.A.; Sensoy, A.; Sorman, A.U. Using MODIS snow cover maps in modeling snowmelt runoff process in the eastern part of Turkey. *Remote Sens. Environ.* **2005**, *97*, 216–230. [CrossRef]
35. Immerzeel, W.W.; Droogers, P.; de Jong, S.M.; Bierkens, M.F.P. Large-scale monitoring of snow cover and runoff simulation in Himalayan river basins using remote sensing. *Remote Sens. Environ.* **2009**, *113*, 40–49. [CrossRef]
36. Finger, D.; Pellicciotti, F.; Konz, M.; Rimkus, S.; Burlando, P. The value of glacier mass balance, satellite snow cover images, and hourly discharge for improving the performance of a physically based distributed hydrological model. *Water Resour. Res.* **2011**, *47*, W07519. [CrossRef]
37. Marti, R.; Gascoin, S.; Berthier, E.; de Pinel, M.; Houet, T.; Laffly, D. Mapping snow depth in open alpine terrain from stereo satellite imagery. *Cryosphere* **2016**, *10*, 1361–1380. [CrossRef]
38. Parajka, J.; Blöschl, G. Validation of MODIS snow cover images over Austria. *Hydrol. Earth Syst. Sci. Discuss.* **2006**, *3*, 1569–1601. [CrossRef]
39. Parajka, J.; Blöschl, G. The value of MODIS snow cover data in validating and calibrating conceptual hydrologic models. *J. Hydrol.* **2008**, *358*, 240–258. [CrossRef]
40. Bavera, D.; De Michele, C. Snow Water Equivalent estimation in Mallero basin using snow gauge data and MODIS images and fieldwork validation. *Hydrol. Process.* **2009**, *23*, 1961–1972. [CrossRef]
41. Bavera, D.; De Michele, C.; Pepe, M.; Rampini, A. Melted snow volume control in the snowmelt runoff model using a snow water equivalent statistically based model. *Hydrol. Process.* **2012**, *26*, 3405–3415. [CrossRef]
42. Bavera, D.; Bavay, M.; Jonas, T.; Lehning, M.; De Michele, C. A comparison between two statistical and a physically-based model in snow water equivalent mapping. *Adv. Water Resour.* **2014**, *63*, 167–178. [CrossRef]
43. Finger, D.; Vis, M.; Huss, M.; Seibert, J. The value of multiple data set calibration versus model complexity for improving the performance of hydrological models in mountain catchments. *Water Resour. Res.* **2015**, *51*. [CrossRef]
44. Şorman, A.A.; Sensoy, A.; Tekeli, A.E.; Sorman, A.U.; Akyurek, Z. modeling and forecasting snowmelt runoff process using the HBV model in the eastern part of Turkey. *Hydrol. Process.* **2009**, *23*, 1031–1040. [CrossRef]
45. Şensoy, A.; Uysal, G. The value of snow depletion forecasting methods towards operational snowmelt runoff estimation using MODIS and Numerical Weather Prediction Data. *Water Resour. Manag.* **2012**, *26*, 3415–3440. [CrossRef]
46. Uysal, G.; Şensoy, A.; Şorman, A.A. Improving daily streamflow forecasts in mountainous Upper Euphrates basin by multi-layer perceptron model with satellite snow products. *J. Hydrol.* **2016**, *543*, 630–650. [CrossRef]
47. Finger, D. The value of satellite retrieved snow cover images to assess water resources and the theoretical hydropower potential in ungauged mountain catchments. *Jökull* **2018**, in press.
48. Lee, S.; Klein, A.G.; Over, T.M. A comparison of MODIS and NOHRSC snowcover products for simulating streamflow using the Snowmelt Runoff Model. *Hydrol. Process.* **2005**, *19*, 2951–2972. [CrossRef]
49. Hall, D.K.; Riggs, G.A. Accuracy assessment of the MODIS snow products. *Hydrol. Process.* **2007**, *21*, 1534–1547. [CrossRef]
50. Da Ronco, P.; de Bárdossy, C. Cloud obstruction and snow cover in Alpine areas from MODIS products. *Hydrol. Earth Syst. Sci.* **2014**, *18*, 4579. [CrossRef]

51. Gafurov, A.; Bárdossy, A. Cloud removal methodology from MODIS snow cover product. *Hydrol. Earth Syst. Sci.* **2009**, *13*, 1361–1373. [CrossRef]

52. Wang, X.; Xie, H. New methods for studying the spatiotemporal variation of snow cover based on combination products of MODIS Terra and Aqua. *J. Hydrol.* **2009**, *371*, 192–200. [CrossRef]

53. Hall, D.K.; Riggs, G.A.; Foster, J.L.; Kumar, S.V. Development and evaluation of a cloud-gap-filled MODIS daily snow-cover product. *Remote Sens. Environ.* **2010**, *114*, 496–503. [CrossRef]

54. Parajka, J.; Pepe, M.; Rampini, A.; Rossi, S.; Blöschl, G. A regional snow-line method for estimating snow cover from MODIS during cloud cover. *J. Hydrol.* **2010**, *381*, 203–212. [CrossRef]

55. Paudel, K.P.; Andersen, P. Monitoring snow cover variability in an agropastoral area in the Trans Himalayan region of Nepal using MODIS data with improved cloud removal methodology. *Remote Sens. Environ.* **2011**, *115*, 1234–1246. [CrossRef]

56. Arslan, A.; Tanis, C.; Metsämäki, S.; Aurela, M.; Böttcher, K.; Linkosalmi, M.; Peltoniemi, M. Automated Webcam Monitoring of Fractional Snow Cover in Northern Boreal Conditions. *Geosciences* **2017**, *7*, 55. [CrossRef]

57. Tanis, C.M.; Peltoniemi, M.; Linkosalmi, M.; Aurela, M.; Böttcher, K.; Manninen, T.; Arslan, A.N. A system for acquisition, processing and visualization of image time series from multiple camera networks. *Data* **2018**, *3*, 23. [CrossRef]

58. Pulliainen, J.T.; Hallikainen, M. Retrieval of regional snow water equivalent from spaceborne passive microwave observations. *Remote Sens. Environ.* **2001**, *75*, 76–85. [CrossRef]

59. Sun, C.; Walker, J.P.; Houser, P.R. A methodology for snow data assimilation in a land surface model. *J. Geophys. Res.* **2004**, *109*, D08108. [CrossRef]

60. Şorman, A.U.; Beser, O. Determination of snow water equivalent over the eastern part of Turkey using passive microwave data. *Hydrol Process.* **2013**, *27*, 1945–1958. [CrossRef]

61. Jörg-Hess, S.; Griessinger, N.; Zappa, M. Probabilistic Forecasts of Snow Water Equivalent and Runoff in Mountainous Areas. *J. Hydrometeorol.* **2015**, *16*, 2169–2186.

62. Slater, A.G.; Clark, M.P. Snow data assimilation via an ensemble Kalman filter. *J. Hydrometeorol.* **2006**, *7*, 478–493. [CrossRef]

63. Foster, J.L.; Sun, C.; Walker, J.P.; Kelly, R.; Chang, A.; Dong, J.; Powell, H. Quantifying the uncertainty in passive microwave snow water equivalent observations. *Remote Sens. Environ.* **2005**, *92*, 187–203. [CrossRef]

64. Dong, J.; Walker, J.P.; Houser, R.P. Factors affecting remotely sensed snow water equivalent uncertainty. *Remote Sens. Environ.* **2005**, *97*, 68–82. [CrossRef]

65. Cordisco, E.; Prigent, C.; Aires, F. Snow characterization at a global scale with passive microwave satellite observations. *J. Geophys. Res.* **2006**, *111*, D19102. [CrossRef]

66. Kelly, R.E.J. The AMSR-E Snow Depth Algorithm: Description and Initial Results. *J. Remote Sens. Soc.* **2009**, *29*, 307–317.

67. Tedesco, M.; Reichle, R.; Loew, A.; Markus, T.; Foster, J.L. Dynamic Approaches for Snow Depth Retrieval from Spaceborne Microwave Brightness Temperature. *IEEE Trans. Geosci. Remote Sens.* **2010**, *48*, 1955–1967. [CrossRef]

68. Tedesco, M.; Narvekar, P. Assessment of the NASA AMSR-E SWE Product. *IEEE J. Sel. Top. Appl. Earth Observ. Remote Sens.* **2010**, *3*, 141–159. [CrossRef]

69. Leppänen, L.; Kontu, A.; Vehviläinen, J.; Lemmetyinen, J.; Pulliainen, J. Comparison of traditional and optical grain-size field measurements with SNOWPACK simulations in a taiga snowpack. *J. Glaciol.* **2015**, *61*, 151–162. [CrossRef]

70. Li, X.; Zhang, L.; Hermüller, L.; Jiang, L.; Vereecken, H. Measurement and Simulation of Topographic Effects on Passive Microwave Remote Sensing Over Mountain Areas. A Case Study from the Tibetan Plateau. *Geosc. Remote Sens.* **2014**, *52*, 1489–1501. [CrossRef]

71. Kontu, A.; Lemmetyinen, J.; Vehviläinen, J.; Leppänen, L.; Pulliainen, J. Coupling SNOWPACK-modeled grain size parameters with the HUT snow emission model. *Remote Sens. Environ.* **2017**, *194*, 33–47. [CrossRef]

72. Conde, V.; Nico, G.; Mateus, P.; Catalão, J.; Kontu, A.; Gritsevich, M. On the estimation of temporal changes of snow water equivalent by spaceborne SAR interferometry: A new application for the Sentinel-1 mission. *J. Hydrol. Hydromech.* **2018**, *67*, 93–100. [CrossRef]

73. Armstrong, R.; Brun, E. *Snow and Climate: Physical Processes, Surface Energy Exchange and Modeling*; Cambridge University Press: Cambridge, UK, 2008.

74. De Michele, C.; Avanzi, F.; Ghezzi, A.; Jommi, C. Investigating the dynamics of bulk snow density in dry and wet conditions using a one-dimensional model. *Cryosphere* **2013**, *7*, 433–444. [CrossRef]

75. Koivusalo, H.; Heikinheimo, M.; Karvonen, T. Test of a simple two–layer parameterisation to simulate the energy balance and temperature of a snowpack. *Theor. Appl. Climatol.* **2001**, *70*, 65–79. [CrossRef]

76. Lehning, M.; Bartelt, P.B.; Brown, R.L.; Fierz, C.; Satyawali, P. A physical SNOWPACK model for the Swiss Avalanche Warning Services. Part II: Snow Microstructure. *Cold Reg. Sci. Technol.* **2002**, *35*, 147–167. [CrossRef]

77. Lehning, M.; Bartelt, P.B.; Brown, R.L.; Fierz, C.; Satyawali, P. A physical SNOWPACK model for the Swiss Avalanche Warning Services. Part III: Meteorological Boundary Conditions, Thin Layer Formation and Evaluation. *Cold Reg. Sci. Technol.* **2002**, *35*, 169–184. [CrossRef]

78. Vionnet, V.; Brun, E.; Morin, S.; Boone, A.; Faroux, S.; le Moigne, P.; Martin, E.; Willemet, J.M. The detailed snowpack 6209 scheme Crocus and its implementation in SURFEX v7.2. *Geosci. Model Dev.* **2012**, *5*, 773–791. [CrossRef]

79. Dutra, E.; Viterbo, P.; Miranda, P.; Balsamo, G. Complexity of Snow Schemes in a Climate Model and Its Impact on Surface Energy and Hydrology. *J. Hydrometeorol.* **2012**, *13*, 521–538. [CrossRef]

80. Best, M.J.; Pryor, M.; Clark, D.B.; Rooney, G.G.; Essery, R.L.H.; Menard, C.B.; Edwards, J.; Hendry, M.A.; Porson, A.; Gedney, N.; et al. The Joint UK Land Environment Simulator (JULES), model description—Part 1: Energy and water fluxes. *Geosci. Model Dev.* **2011**, *4*, 677–699. [CrossRef]

81. Boone, A. *Description du Schema de Neige ISBA-ES (Explicit Snow)*; Centre National de Recherches: Toulouse, France, 2002.

82. Zängl, G.; Reinert, D.; Ripodas, P. The ICON (ICOsahedral Non-hydrostatic) modeling framework of DWD and MPI-M: Description of the non-hydrostatic dynamical core. *Q. J. R. Meteorol. Soc.* **2014**, *141*, 563–579. [CrossRef]

83. Carmagnola, C.M.; Morin, S.; Lafaysse, M.; Domine, F.; Lesaffre, B.; Lejeune, Y.; Picard, G.; Arnaud, L. Implementation and evaluation of prognostic representations of the optical diameter of snow in the SURFEX/ISBA-Crocus detailed snowpack model. *Cryosphere* **2014**, *8*, 417–437. [CrossRef]

84. Avanzi, F.; De Michele, C.; Ghezzi, A.; Jommi, C.; Pepe, M. A processing-modeling routine to use SNOTEL hourly data in snowpack dynamic models. *Adv. Water Resour.* **2014**, *73*, 16–29. [CrossRef]

85. Slater, A.G.; Schlosser, C.A.; Desborough, C.E.; Pitman, A.J.; Henderson-Sellers, A.; Robock, A.; Vinnikov, K.Y.; Mitchell, K.; Boone, A.; Braden, H.; et al. The representation of snow in land-surface schemes: Results from PILPS 2(d). *J. Hydrometeorol.* **2001**, *2*, 7–25. [CrossRef]

86. Rutter, N.; Essery, R.; Pomeroy, J.; Altimir, N.; Andreadis, K.; Baker, I.; Yamazaki, T. Evaluation of forest snow processes models (SnowMIP2). *J. Geophys. Res.* **2009**, *114*, 18. [CrossRef]

87. Dutra, E.; Balsamo, G.; Viterbo, P.; Miranda, P.M.A.; Beljaars, A.C.M.; Schär, C.; Elder, K. An improved snow scheme for the ECMWF land surface model: Description and offline validation. *J. Hydrometeor.* **2010**, *11*, 899–916. [CrossRef]

88. Andreadis, K.M.; Lettenmaier, D.P. Assimilating remotely sensed snow observations into a macroscale hydrology model. *Adv. Water Resour.* **2006**, *29*, 872–886. [CrossRef]

89. Clark, M.P.; Slater, A.G.; Barrett, A.P.; Hay, L.E.; Mccabe, G.J.; Rajagopalan, B.; Leavesley, G.H. Assimilation of snow covered area information into hydrologic and land-surface models and land-surface models. *Adv. Water Resour.* **2006**, *29*, 1209–1221. [CrossRef]

90. Leisenring, M.; Moradkhani, H. Snow water equivalent prediction using Bayesian data assimilation methods. *Stoch. Environ. Res. Risk A* **2011**, *25*, 253–270. [CrossRef]

91. Nagler, T.; Rott, H.; Malcher, P.; Müller, F. Assimilation of meteorological and remote sensing data for snowmelt runoff forecasting. *Remote Sens. Environ.* **2008**, *112*, 1408–1420. [CrossRef]

92. Liu, Y.; Peters-Lidard, C.D.; Kumar, S.; Foster, J.L.; Shaw, M.; Tian, Y.; Fall, G.M. Assimilating satellite-based snow depth and snow cover products for improving snow predictions in Alaska. *Adv. Water Resour.* **2013**, *54*, 208–227. [CrossRef]

93. Saloranta, T.M. Operational snow mapping with simplified data assimilation using the seNorge snow model. *J. Hydrol.* **2016**, *538*, 314–325. [CrossRef]

94. Takala, M.; Ikonen, J.; Luojus, K.; Lemmetyinen, J.; Metsämäki, S.; Cohen, J.; Arslan, A.; Pulliainen, J. New Snow Water Equivalent Processing System with Improved Resolution Over Europe and its Applications in Hydrology. *IEEE J. Sel. Top. Appl. Earth Observ. Remote Sens.* **2017**, *10*, 428–436. [CrossRef]

95. Fletcher, S.J.; Liston, G.E.; Hiemstra, C.A.; Miller, S.D. Assimilating MODIS and AMSR-E snow observations in a snow evolution model. *J. Hydrometeorol.* **2012**, *13*, 1475–1492. [CrossRef]

96. Bergeron, J.M.; Trudel, M.; Leconte, R. Combined assimilation of streamflow and snow water equivalent for mid-term ensemble streamflow forecasts in snow-dominated regions. *Hydrol. Earth Syst. Sci.* **2016**, *20*, 4375–4389. [CrossRef]

97. Charrois, L.; Cosme, E.; Dumont, M.; Lafaysse, M.; Morin, S.; Libois, Q.; Picard, G. On the assimilation of optical reflectances and snow depth observations into a detailed snowpack model. *Cryosphere* **2016**, *10*, 1021–1038. [CrossRef]

98. Dziubanski, D.J.; Franz, K.J. Assimilation of AMSR-E snow water equivalent data in a spatially-lumped snow model. *J. Hydrol.* **2016**, *540*, 26–39. [CrossRef]

99. Griessinger, N.; Seibert, J.; Magnusson, J.; Jonas, T. Assessing the benefit of snow data assimilation for runoff modeling www.hydrol-earth-syst-sci.net/21/635/2017/. *Hydrol. Earth Syst. Sci.* **2017**, *21*, 635–650.

100. Alvarado-Montero, R.; Schwanenberg, D.; Krahe, P.; Lisniak, D.; Sensoy, A.; Sorman, A.; Akkol, B. Moving Horizon Estimation for Assimilating H-SAF Remote Sensing Data into the HBV Hydrological Model. *Adv. Water Resour.* **2016**, *92*, 248–257. [CrossRef]

101. Huang, C.; Newman, A.J.; Clark, M.P.; Wood, A.W.; Zheng, X. Evaluation of snow data assimilation using the ensemble Kalman filter for seasonal streamflow prediction in the western United States. *Hydrol. Earth Syst. Sci.* **2017**, *21*, 635–650. [CrossRef]

102. Piazzi, G.; Thirel, G.; Campo, L.; Gabellani, S. A particle filter scheme for multivariate data assimilation into a point-scale snowpack model in an Alpine environment. *Cryosphere* **2018**, *12*, 2287–2306. [CrossRef]

103. Liston, G.; Hiemstra, C.A. A simple data assimilation system for complex snow distributions (SnowAssim). *J. Hydrometeorol.* **2008**, *9*, 989–1004. [CrossRef]

104. Liston, G.E.; Pielke, R.A., Sr.; Greene, E.M. Improving first-order snow-related deficiencies in a regional climate model. *J. Geophys. Res.* **1999**, *104*, 19559–51567. [CrossRef]

105. Houser, P.R.; De Lannoy, G.; Walker, J.P. Land Surface Data Assimilation, p549-598. In *Data Assimilation: Making Sense of Observations*; Lahoz, W., Khatattov, B., Menard, R., Eds.; Springer: Dordrecht, The Netherlands, 2010; p. 732.

106. Barrett, A.P. National operational hydrologic remote sensing center snow data assimilation system (SNODAS) products at NSIDC. Special Rep. 11, NSIDC: Boulder, CO, USA, 2003; p. 19. Available online: https://nsidc.org/pubs/documents/special/nsidc_special_report_11.pdf (accessed on 14 December 2018).

107. Brasnett, B. A global analysis of snow depth for numerical weather prediction. *J. App. Meteorol.* **1999**, *38*, 726–740. [CrossRef]

108. Rodell, M.; Houser, P.R. Updating a land surface model with MODIS-derived snow cover. *J. Hydrometeorol.* **2004**, *5*, 1064–1075. [CrossRef]

109. Zaitchik, B.F.; Rodell, M. Forward-looking assimilation of MODIS-derived snow-covered area into a land surface model. *J. Hydrometeorol.* **2009**, *10*, 130–148. [CrossRef]

110. Dong, J.; Walker, J.; Houser, P.; Sun, C. Scanning multichannel microwave radiometer snow water equivalent assimilation. *J. Geophys. Res.* **2007**, *112*, D07108. [CrossRef]

111. Durand, M.; Margulis, S.A. Feasibility test of multifrequency radiometric data assimilation to estimate snow water equivalent. *J. Hydrometeorol.* **2006**, *7*, 443–457. [CrossRef]

112. Durand, M.; Margulis, S.A. Correcting first-order errors in snow water equivalent estimates using a multifrequency, multiscale radiometric data assimilation scheme. *J. Geophys. Res.* **2007**, *112*, D13. [CrossRef]

113. Andreadis, K.M.; Liang, D.; Tsang, L.; Lettenmaier, D.P.; Josberger, E.G. Characterization of errors in a coupled snow hydrology—microwave emission model. *J. Hydrometeorol.* **2008**, *9*, 149–164. [CrossRef]

114. Durand, M.; Kim, E.J.; Margulis, S.A. Radiance assimilation shows promise for snowpack characterization. *Geophys. Res. Lett.* **2009**, *36*. [CrossRef]

115. Che, T.; Li, X.; Jin, R.; Huang, C. Assimilating passive microwave remote sensing data into a land surface model to improve the estimation of snow depth. *Remote Sens. Environ.* **2014**, *143*, 54–63. [CrossRef]

116. Li, D.; Durand, M.; Margulis, S. Estimating snow water equivalent in a Sierra Nevada watershed via spaceborne radiance data assimilation. *Water Resour. Res.* **2017**, *53*. [CrossRef]

117. Larue, F.; Royer, A.; De Sève, D.; Roy, A.; Picard, G.; Vionnet, V. Simulation and assimilation of passive microwave data using a snowpack model coupled to a calibrated radiative transfer model over northeastern Canada. *Water Resour. Res.* **2018**, *54*, 4823–4848. [CrossRef]

118. Larue, F.; Royer, A.; De Sève, D.; Roy, A.; Cosme, E. Assimilation of passive microwave AMSR-2 satellite observations in a snowpack evolution model over North-Eastern Canada. *Hydrol. Earth Syst. Sci. Discuss.* **2018**. under review. [CrossRef]

119. Kwon, Y.; Toure, A.M.; Yang, Z.-L.; Rodell, M.; Picard, G. Error characterization of the coupled land surface–radiative transfer models for snow passive microwave radiance assimilation. *IEEE Trans. Geosci. Remote Sens.* **2015**, *53*, 5247–5268. [CrossRef]

120. Lemmetyinen, J.; Pulliainen, J.; Rees, A.; Kontu, A.; Qiu, Y.; Derksen, C. Multiple-Layer Adaptation of HUT Snow Emission Model. Comparison with Experimental Data. *IEEE Tran. Geosci. Remote Sens.* **2010**, *48*, 2781–2794. [CrossRef]

121. Wiesmann, A.; Mätzler, C. Microwave emission model of layered snowpacks. *Remote Sens. Environ.* **1999**, *70*, 307–316. [CrossRef]

122. Tsang, L.; Pan, J.; Liang, D.; Li, Z.; Cline, D. Modeling Active Microwave Remote Sensing of Snow using Dense Media Radiative Transfer (DMRT) Theory with Multiple Scattering Effects. *IEEE Int. Symp. Geosci. Remote Sens.* **2006**. [CrossRef]

123. Picard, G.; Brucker, L.; Roy, A.; Dupont, F.; Fily, M.; Royer, A. Simulation of the microwave emission of multi-layered snowpacks using the dense media radiative transfer theory, the DMRT-ML model. *Geosci. Model Dev.* **2013**. [CrossRef]

124. Royer, A.; Roy, A.; Montpetit, B.; Saint-Jean-Rondeau, O.; Picard, G.; Brucker, L.; Langlois, A. Comparison of commonly-used microwave radiative transfer models for snow remote sensing. *Remote Sens. Environ.* **2017**, *190*, 247–259. [CrossRef]

125. Löwe, H.; Picard, G. Microwave scattering coefficients of snow in MEMLS and DMRT-ML revisited: The relevance of sticky hard spheres and tomography-based estimates of stickiness. *Cryosphere* **2015**, *9*, 2101–2117. [CrossRef]

126. Pan, J.; Durand, M.; Sandells, M.; Lemmetyinen, J.; Kim, E.J.; Pulliainen, J. Differences between the HUT Snow Emission Model and MEMLS and Their Effects on Brightness Temperature Simulation. *IEEE Trans. Geosci. Remote Sens.* **2016**, *54*, 2001–2019. [CrossRef]

127. Picard, G.; Sandells, M.; Löwe, H. SMRT: An active–Passive microwave radiative transfer model for snow with multiple microstructure and scattering formulations (v1.0). *Geosci. Model Dev.* **2018**, *11*, 2763–2788. [CrossRef]

128. Sandells, M.; Essery, R.; Rutter, N.; Wake, L.; Leppänen, L.; Lemmetyinen, J. Microstructure representation of snow in coupled snowpack and microwave emission models. *Cryosphere* **2017**, *11*, 229–246. [CrossRef]

129. COST ESSEM 1404, Memorandum of Understanding, Brussels, 15 May, 2015, COST 032/14. Available online: https://e-services.cost.eu/files/domain_files/ESSEM/Action_ES1404/mou/ES1404-e.pdf (accessed on 13 December 2018).

130. Malik, N.; Bookhagen, B.; Marwan, N.; Kurths, J. Analysis of spatial and temporal extreme monsoonal rainfall over South Asia using complex networks. *Clim. Dyn.* **2012**, *39*, 971–987. [CrossRef]

131. Cressman, G.P. An operational objective analysis system. *Mon. Weather Rev.* **1959**, *87*, 367–374. [CrossRef]

132. Dee, D.; Uppala, S.; Simmons, A.; Berrisford, P.; Poli, P.; Kobayashi, S.; Andrae, U.; Balsameda, M.; Balsamo, G.; Bauer, P.; et al. The ERA-Interim reanalysis: Configuration and performance of the data assimilation system. *Q. J. R. Meteorol. Soc.* **2011**, *137*, 553–597. [CrossRef]

133. De Rosnay, P.; Balsamo, G.; Albergel, C.; Muñoz-Sabater, J.; Isaksen, L. Initialisation of land surface variables for Numerical Weather Prediction. *Surv. Geophys.* **2014**, *35*, 607–621. [CrossRef]

134. De Rosnay, P.; Isaksen, L.; Dahoui, M. Snow data assimilation at ECMWF. *ECMWF Newslett.* **2015**, *143*, 26–31.

135. Stauffer, D.R.; Seaman, N.L. Use of four-dimensional data assimilation in a limited-area mesoscale model. Part I: Experiments with synoptic-scale data. *Mon. Weather Rev.* **1990**, *118*, 1250–1277. [CrossRef]

136. Boni, G.; Castelli, F.; Gabellani, S.; Machiavello, G.; Rudari, R. Assimilation of MODIS snow cover and real time snow depth point data in a snow dynamic model. *Geosci. Remote Sens. Symp.* **2010**, 1788–1791. [CrossRef]

137. Kalman, R.E. A new approach to linear filtering and prediction problems. *J. Basic Eng.* **1960**, *82*, 35–45. [CrossRef]

138. Gelb, A. Optimal linear filtering. In *Applied Optimal Estimation*; MIT Press: Cambridge, MA, USA, 1974; pp. 102–155.

139. Miller, R.N.; Ghil, M.; Gauthiez, F. Advanced data assimilation in strongly nonlinear dynamical systems. *J. Atmos. Sci.* **1994**, *51*, 1037–1056. [CrossRef]

140. Moradkhani, H. Hydrologic remote sensing and land surface data assimilation. *Sensors* **2008**, *8*, 2986–3004. [CrossRef]

141. Evensen, G. Sequential data assimilation with a nonlinear quasi-geostrophic model using Monte Carlo methods to forecast error statistics. *J. Geophys. Res.* **1994**, *99*, 10143–10162. [CrossRef]

142. Evensen, G. The Ensemble Kalman Filter: Theoretical formulation and practical implementation. *Ocean Dynam.* **2003**, *53*, 343–367. [CrossRef]

143. Arulampalam, M.S.; Maskell, S.; Gordon, N.; Clapp, T. A tutorial on particle filters for on-line non-linear/non-Gausssian Bayesin tracking. *IEEE Trans. Signal Process.* **2002**, *50*, 174–188. [CrossRef]

144. Moradkhani, H.; Sorooshian, S.; Gupta, H.V.; Houser, P.R. Dual state–Parameter estimation of hydrological models using ensemble Kalman filter. *Adv. Water Resour.* **2005**, *28*, 135–147. [CrossRef]

145. Zhou, Y.; McLaughlin, D.; Entekhabi, D. Assessing the performance of the ensemble Kalman filter for land surface data assimilation. *Mon. Wea. Rev.* **2006**, *134*, 2128–2142. [CrossRef]

146. Moradkhani, H.; Sorooshian, S. General review of rainfall-runoff modeling, model calibration, data assimilation, and uncertainty analysis. *Hydrol. Model. Water Cycle* **2009**, *63*, 1–24.

147. Montzka, C.; Moradkhani, H.; Weihermuller, L.; Canty, M.; Hendricks Franssen, H.J.; Vereecken, H. Hydraulic Parameter Estimation by Remotely-sensed top Soil Moisture Observations with the Particle Filter. *J. Hydrol.* **2011**, *399*, 410–421. [CrossRef]

148. Bocquet, M.; Pires, C.A.; Wu, L. Beyond Gaussian statistical modeling in geophysical data assimilation (Review). *Mon. Weather Rev.* **2010**, *138*, 2997–3023. [CrossRef]

149. Li, Z.; Navon, I.M. Optimality of variational data assimilation and its relationship with the Kalman filter and smoother. *Q. J. R. Meteorol. Soc.* **2001**, *127*, 661–683. [CrossRef]

150. Allgöwer, F.; Badgwell, T.A.; Qin, J.S.; Rawlings, J.B.; Wright, S.J. Nonlinear Predictive Control and Moving Horizon Estimation An Introductory Overview. In *Advances in Control, Highlights of ECC99*; Frank, P.M., Ed.; Springer Verlag: Berlin, Germany, 1999; pp. 391–449.

151. Alvarado-Montero, R.; Schwanenberg, D.; Krahe, P.; Helmke, P.; Klein, B. Multi-parametric variational data assimilation for hydrological forecasting. *Adv. Water Resour.* **2017**, *110*, 182–192. [CrossRef]

152. Su, H.; Yang, Z.L.; Niu, G.Y.; Dickinson, R.E. Enhancing the estimation of continental-scale snow water equivalent by assimilating MODIS snow cover with the ensemble Kalman filter. *J. Geophys. Res. Atmos.* **2008**, *113*, D08120. [CrossRef]

153. Kumar, S.V.; Reichle, R.H.; Peters-Lidard, C.D.; Koster, R.D.; Zhan, X.; Crow, W.T.; Eylander, J.B.; Houser, P.R. A land surface data assimilation framework using the land information system: Description and applications. *Adv. Water Resour.* **2008**, *31*, 1419–1432. [CrossRef]

154. Durand, Y.; Laternser, M.; Giraud, G.; Etchevers, P.; Lesaffre, L.; Mérindol, L. Reanalysis of 44 year of climate in the French Alps (1958–2002): Methodology, model validation, climatology, and trends for air temperature and precipitation. *J. Appl. Meteorol. Clim.* **2009**, *48*, 29–449.

155. Durand, Y.; Laternser, M.; Giraud, G.; Etchevers, P.; Mérindol, L.; Lesaffre, B. Reanalysis of 47 Years of Climate in the French Alps (1958–2005): Climatology and Trends for Snow Cover. *J. Appl. Meteorol. Clim.* **2009**, *48*, 2487–2512. [CrossRef]

156. Toure, A.M.; Goïta, K.; Royer, A.; Kim, E.J.; Durand, M.; Margulis, S.A.; Lu, H. A case study of using a multilayered thermodynamical snow model for radiance assimilation. *IEEE Trans. Geosci. Remote Sens.* **2011**, *49*, 2828–2837. [CrossRef]

157. Durand, M.; Margulis, S.A. Effects of uncertainty magnitude and accuracy on assimilation of multi-scale measurements for snowpack characterization. *J. Geophys. Res. Atmos.* **2008**, *113*, D02105. [CrossRef]

158. Su, H.; Yang, Z.-L.; Dickinson, R.E.; Wilson, C.R.; Niu, G.-Y. Multisensor snow data assimilation at the continental scale: The value of Gravity Recovery and Climate Experiment terrestrial water storage information. *J. Geophys. Res.* **2010**, *115*, D10104. [CrossRef]

159. Magnusson, J.; Gustafsson, D.; Hüsler, F.; Jonas, T. Assimilation of point SWE data into a distributed snow cover model comparing two contrasting methods. *Water Resour. Res.* **2014**, *50*, 7816–7835. [CrossRef]

160. Griessinger, N.; Seibert, J.; Magnusson, J.; Jonas, T. Evaluation of snow data assimilation in Alpine catchments. *Hydrol. Earth Syst. Sci.* **2016**, *20*, 3895–3905. [CrossRef]

161. Magnusson, J.; Winstral, A.; Stordal, A.S.; Essery, R.; Jonas, T. Improving physically based snow simulations by assimilating snow depths using the particle filter. *Water Resour. Res.* **2017**, *53*, 1125–1143. [CrossRef]

162. Pullen, S.; Jones, C.; Rooney, G. Using satellite-derived snow cover data to implement a snow analysis in the met office NWP model. *J. Appl. Meteorol.* **2011**, *50*, 958–973. [CrossRef]

163. Del Moral, P. Non Linear Filtering: Interacting Particle Solution. *Markov Process. Relat. Fields* **1996**, *2*, 555–580.

164. Rood, R.B.; Cohn, S.E.; Coy, L. Data assimilation for EOS: The value of assimilated data. Part 1. *Earth Obs.* **1994**, *6*, 23–25.

165. Walker, J.P.; Houser, P.R. Hydrologic data assimilation. In *Advances in Water Science Methodologies*; Balkema: Rotterdam, The Netherlands, 2005; p. 230.

166. World Meteorological Organization (WMO). *Manual on the Global Observing System, Volume I—Global aspects: Annex V to the WMO Technical Regulations*; (2015 edition, updated in 2017), WMO- No. 544; WMO: Geneva, The Switzerland, 2015; ISBN 978-92-63-10544-8.

167. Workshop Report 1st Snow Data Assimilation Workshop in the framework of COST HarmoSnow ESSEM 1404. Available online: https://www.schweizerbart.de/papers/metz/detail/prepub/89726/Workshop_Report_1st_Snow_Data_Assimilation_Workshop_in_the_framework_of_COST_HarmoSnow_ESSEM_1404 (accessed on 13 December 2018).

168. Ramsay, B. The interactive multisensor snow and ice mapping system. *Hydrol. Process.* **1998**, *12*, 1537–1546. [CrossRef]

169. Kongoli, C.; Dean, C.; Helfrich, S.; Ferraro, R. Evaluating the potential of a blended passive microwave-interactive multi-sensor product for improved mapping of snow cover and estimations of snow water equivalent. *Hydrol. Process.* **2007**, *21*, 1597–1607. [CrossRef]

170. Gao, Y.; Xie, H.J.; Lu, N.; Yao, T.D.; Liang, T.G. Toward advanced daily cloud-free snow cover and snow water equivalent products from Terra-Aqua MODIS and Aqua AMSR-E measurements. *J. Hydrol.* **2010**, *385*, 23–35. [CrossRef]

171. Akyurek, Z.; Hall, D.K.; Riggs, G.A.; Sorman, A.U. Evaluating the utility of the ANSA blended snow cover product in the mountains of eastern Turkey. *Int. J. Remote Sens.* **2010**, *31*, 3727–3744. [CrossRef]

172. Foster, J.L.; Hall, D.K.; Eylander, J.B.; Riggs, G.A.; Nghiem, S.V.; Tedesco, M.; Kim, E.J.; Montesano, P.M.; Kelly, R.E.J.; Casey, K.A.; et al. A blended global snow product using visible, passive microwave and scatterometer data. *Int. J. Remote Sens.* **2011**, *32*, 1371–1395. [CrossRef]

173. Janjić, T.; Bormann, N.; Bocquet, M.; Carton, J.A.; Cohn, S.E.; Dance, S.L.; Losa, S.N.; Nichols, N.K.; Potthast, R.; Waller, J.A.; et al. On the representation error in data assimilation. *Q. J. R. Meteorol. Soc.* **2017**, *144*, 713. [CrossRef]

174. Kurzeneva, E.; Choulga, M.; Rontu, L. Error Statistics in Data Assimilation for NWP: Perspectives for Snow. In Proceedings of the Workshop: Towards a Better Harmonization of Snow Observations, Modeling and Data Assimilation in Europe, Budapest, Hungary, 30–31 October 2018.

175. Kumar, S.V.; Peters-Lidard, C.D.; Arsenault, K.R.; Getirana, A.; Mocko, D. Quantifying the added value of snow cover area observations in passive microwave snow depth assimilation. *J. Hydrometeor.* **2015**, *16*, 1736–1741. [CrossRef]

176. He, M. Data Assimilation in Watershed Models for Improved Hydrologic Forecasting. Ph.D Thesis, University of California, Los Angeles, CA, USA, 2010; p. 173.

177. He, M.; Hogue, T.S.; Franz, K.J.; Margulis, S.A. An integrated uncertainty and ensemble-based data assimilation framework for improved operational streamflow predictions. *Hydrol. Earth Syst. Sci.* **2012**, *16*, 815–831. [CrossRef]

178. Franz, K.J.; Hogue, T.S.; Barik, M.; He, M. Assessment of SWE data assimilation for ensemble streamflow predictions. *J. Hydrol.* **2014**, *519*, 2737–2746. [CrossRef]

179. COST ESSEM 1404, Training School on Snow Observations and Data Assimilation in Bormio, 12–16 March 2018. Available online: http://www.harmosnow.eu/index.php?page=Training%20School%20Bormio (accessed on 13 December 2018).

Article

Generating Observation-Based Snow Depletion Curves for Use in Snow Cover Data Assimilation

Kristi R. Arsenault [1,*] **and Paul R. Houser** [2]

[1]   SAIC, Inc., McLean, VA 22101, USA

[2]   Department of Geography and Geoinformation Science, George Mason University, Fairfax, VA 22030, USA; phouser@gmu.edu

*   Correspondence: kristi.r.arsenault@nasa.gov; Tel.: +1-301-614-5772

Received: 1 November 2018; Accepted: 10 December 2018; Published: 14 December 2018

Abstract: Snow depletion curves (SDC) are functions that are used to show the relationship between snow covered area and snow depth or water equivalent. Previous snow cover data assimilation (DA) studies have used theoretical SDC models as observation operators to map snow depth to snow cover fraction (SCF). In this study, a new approach is introduced that uses snow water equivalent (SWE) observations and satellite-based SCF retrievals to derive SDC relationships for use in an Ensemble Kalman filter (EnKF) to assimilate snow cover estimates. A histogram analysis is used to bin the SWE observations, which the corresponding SCF observations are then averaged within, helping to constrain the amount of data dispersion across different temporal and regional conditions. Logarithmic functions are linearly regressed with the binned average values, for two U.S. mountainous states: Colorado and Washington. The SDC-based logarithmic functions are used as EnKF observation operators, and the satellite-based SCF estimates are assimilated into a land surface model. Assimilating satellite-based SCF estimates with the observation-based SDC shows a reduction in SWE-related RMSE values compared to the model-based SDC functions. In addition, observation-based SDC functions were derived for different intra-annual and physiographic conditions, and landcover and elevation bands. Lower SWE-based RMSE values are also found with many of these categorical observation-based SDC EnKF experiments. All assimilation experiments perform better than the open-loop runs, except for the Washington region's 2004–2005 snow season, which was a major drought year that was difficult to capture with the ensembles and observations.

Keywords: snow cover; snow depletion curve; MODIS; data assimilation; land surface model

1. Introduction

Snow depletion curves (SDC) define the relationship between changes in snow cover area (SCA) and the snow pack, which can impact, for example, snow-albedo feedback in global climate models [1] and the amount of water storage and melt for hydrological models [2,3]. SDCs typically involve functions that are fit or tuned to a given set of snow-based observations or theoretical conditions. Currently, many hydrological, land surface models (LSMs), and climate models use simple to complex schemes to define this snow depth-cover relationship, with many models still using very simple schemes, which do not account for even regional or temporal changes [4]. Some approaches to estimating snow depth-snow cover relationships involve statistical approaches, from using simple fitted functions to more shape and scale parameter-based gamma and beta distributions [4–8]. Much of the snowpack depletion in mountainous regions relates to late winter and early spring peak snow water equivalent (SWE) melt energy and radiation spatial variations (e.g., [2,9]), even though SWE can decrease without decreasing areal snow cover.

Several SDC schemes have been used to map the predicted SWE or snow depth states to snow cover fraction (SCF) (or vice versa) for different snow cover data assimilation (DA) approaches [10–12].

The SDC scheme can be used as the observation operator, which relates the model variables to the observations. In this case, the SDC acts as the observation operator and converts snow-based model estimates (e.g., SWE) to be in the same units and similar value range as the snow cover observation estimates (e.g., snow cover fraction), for calculating the innovation and assimilation update. The innovation step is simply the difference taken between model-generated snow cover and the observed snow cover estimates. Some studies have utilized the Ensemble Kalman Filter (EnKF) to assimilate snow cover fraction or area observations [10,13,14], as it has been shown to perform overall better than simpler methods, e.g., direct insertion [15]. EnKF relies on an ensemble of model forecasts generated using error covariances. Snow cover assimilation using the EnKF method has been applied at different scales, including global [16], continental [13], and regional [10,14,15,17]. Other ensemble-based snow cover DA studies have used the ensemble square-root filter [18] and the particle filter [19].

A variety of SDC functions have been used as the observation operator in these past snow-cover DA studies. Cumulative density functions (CDFs) of beta distributions for varying conditions have been applied regionally [10], or tuning of a hyperbolic tangent formulation, using shape parameters and a snow density parameter [20], has been applied also for different scales [13,16]. Other studies have used, for example, a two-parameter based lognormal probability distribution function, where snow cover area was represented as a summation of areal SWE and the SDCs were tuned by varying different coefficient of variation (CV) parameters for melt and accumulation periods [18]. Such SDC-based observation operator schemes have accounted for varying conditions, such as elevation, vegetation and accumulation and ablation phases [10,13,17]. Accounting for the accumulation and ablation phases, also referred to as the hysteresis characteristic of snow, can be a very important aspect of the SDC scheme, since different curves can represent these phases and are reflected in the curve parameters (e.g., [10,17]).

Previous studies that have assimilated satellite-based snow cover to improve model snow states and other hydrological variables (e.g., streamflow) have utilized snow observations to tune the snow depletion curves. Despite these previous efforts, many of the SDCs used in snow cover assimilation may have been considered theoretically simplistic [12,14] or tuned with very coarse spatial scale snow observations [20], and these SDCs may not be suitable enough for assimilating snow cover data for finer scale, mountainous regions. Also, what if the snow observations themselves were used to derive the snow depletion curves to be used as the observation operators for snow cover assimilation? One recent study used satellite-based snow cover and five snow depth stations to derive SDC functions, based on the hyperbolic tangent formulation, to derive such observation operators for a mountainous catchment in China [17]. Xu et al. [17] generated curves separately for each station and the two snow phases, accumulation and ablation. Assimilating snow cover into an LSM using these highly tuned observational based observation operators improved the modeled snow depth states at those few sites [17]. Another recent study, which focused on generating SDCs at different scales by using observed SWE and satellite-based snow cover area, showed how using such observational information can help estimate better SWE conditions (e.g., peak SWE) when 100% snow cover occurs (for an area or grid cell), given the location or area being represented [2].

In this study, we present a new approach to estimating snow depletion curves and their application for assimilating snow cover fraction observations, using an EnKF data assimilation approach and a land surface model with a multi-layer snow physics scheme. Building upon these recent studies, we use observed SWE and snow cover fraction estimates to derive new SDCs, using a larger array of observations, spanning two different mountainous regions in the United States (U.S.). We refer to these new SDC-type observation operators as "observation-based" and benchmark their skill against the default, model-based SDC function. These new SDC observation-based observation operators are hypothesized to improve the model-based SCF forecasts and snow state analysis. A secondary goal in applying this SDC approach is to see how accounting for varying vegetation, elevation, and temporal conditions may better capture heterogeneous features related to the snowpack and snow cover patterns when assimilating snow cover observations. Finally, the new SDC-based observation operators are used to derive the observational errors, which are used in the EnKF method. Accounting for different errors related to the varying conditions provides different weighting of the snow cover observations against that of the model in the EnKF innovation and update steps.

The primary research question is: How do these new observation-based, SDC-type operational operators perform relative to a default model-based scheme when assimilating snow cover estimates? The secondary research question is: How does accounting for different temporal and physiographic conditions in relation to the new observation-based SDC impact the assimilation of the snow cover estimates? We address these questions throughout the paper within the following sections. Section 2 provides an overview of the study regions, and Section 3 provides several details of the snow observations, the multi-layer snow model, and data assimilation method used in this study. Section 4 describes the method on how the new observation-based SDCs are derived and applied in the EnKF assimilation approach, and Section 5 presents the results of the new SDCs applied as the observation operator when assimilating the snow cover observations. Finally, we discuss the results, benefits and challenges of this new SDC approach in the Discussion and Conclusions section.

2. Study Area

The study domains span the mountainous regions of Washington (WA) and Colorado (CO) states within the United States. The spatial grid is on an equirectangular projection with a resolution of 0.01 degree (~1 km) with bounding coordinates of the lower left corner (45.005° N, −121.995° W) and upper right corner (48.995° N, −116.995° W) for WA (400 latitudinal and 501 longitudinal points), and lower left corner (37.005° N, −108.995° W) and upper right corner (40.995° N, −102.005° W) for CO (400 latitudinal and 700 longitudinal points). The 0.01° gridcells are represented at the center of the gridcell, as reflected in the bounding coordinates. Figure 1 highlights the WA and CO study areas, depicting the areas' high topographic variability (e.g., mountainous and low-land areas). The 0.01° elevation maps are derived from the 30 m U.S. Geological Survey (USGS) National Elevation Dataset (NED; https://lta.cr.usgs.gov/NED). Also, these areas have a wide range of soil and vegetation types (e.g., coniferous forests, grasslands, agricultural, etc.). The regions were selected to capture not only the topographic variability, but also variability in surface snow amounts. Figure 1 also shows each region's associated in-situ U.S. Department of Agriculture's (USDA) Natural Resources Conservation Service (NRCS) SNOwpack TELemetry (SNOTEL) measurement networks used in this study. After additional quality-control checks, 56 SNOTEL stations are identified as being within the WA domain, and 98 SNOTEL stations are within the CO domain (represented by black triangles). These stations are used in deriving and the new SDCs, as described in Section 4.1.

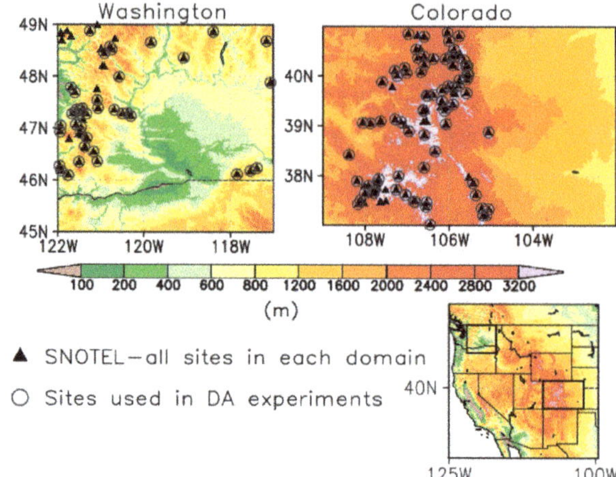

Figure 1. Elevation maps of Washington and Colorado domains, from the U.S. Geological Survey (USGS) National Elevation Dataset (NED) dataset. All SNOwpack TELemetry (SNOTEL) site (black triangles) are used in generating the observation-based snow depletion curves (SDCs), and a reduced amount of site locations are used only in the model experiments and validation (grey open circles).

3. Data and Model Sources

3.1. Observational Datasets Used

The Terra MODIS Level 3, collection 5 (MOD10A1), daily 500 m snow cover fraction (SCF) product has been selected for the work presented here [21,22]. This product is derived from the MODIS Normalized Difference Snow Index (NDSI) to estimate an SCF value for each 500 m pixel [21, 23]) The MODIS snow cover products are obtained from the National Snow and Ice Data Center (NSIDC) in Boulder, Colorado, USA (http://nsidc.org/data/mod10a1) [24]. The 500 m MODIS snow cover estimates have been shown to be sufficient in high mountain terrain [25]. The daily 500 m products come in a sinusoidal projection, which are resampled with the nearest neighbor using the MODIS Reprojection Tool [26] to a 0.01° geographic coordinate system (GCS) for our model and DA experiments. The original 500 m sinusoidal projection dataset is used for deriving the observation-based SDCs and observational errors, which are further described in Section 4.

The main SWE observation dataset used in this study is the SNOTEL network dataset [27]. The daily SNOTEL SWE measurements are observed using snow pillow instruments. The SNOTEL network is sometimes one of the only datasets available in mountainous areas in the U.S. to validate model estimates [28], which provide important information, like streamflow, in key water supply areas in the mountainous West [29]. The SNOTEL data used in this study include Water Years 2000–2010, which overlap the Terra/MODIS data period selected. Note: A water year (WY) is used in many hydrological studies and starts on October 1 and goes to September 30 of the following year, spanning one full snow season cycle. The SNOTEL stations are point measurements, but applied here as a representative over the gridcell (e.g., 0.01° cells). New SDCs are derived from the above MODIS SCF and SNOTEL SWE observations (which is further described in Section 4).

3.2. Land Surface Model Description and Setup

The LSM used in this study is the Community Land Model, version 2.0 (CLM2) [30–32]. CLM2 is run in an off-line mode, uncoupled to any atmospheric or other modeling component, and driven within the Land Information System (LIS) framework [33]. The use of LIS allows us to take advantage of its many DA, observational, and model interface capabilities. At the time of this study, later versions of CLM were not available in LIS though CLM2 was already implemented and well tested. The model contains ten soil layers with varying thickness, and a five-layer snow scheme, which accounts for liquid, ice, and heat energy storages and changes within each layer. The water balance equations are mass conserving with inputs provided mainly by precipitation and outputs associated through evapotranspiration sources (e.g., vegetation and soil) and runoff. The snow parameterizations are adapted mostly from other studies [34–36]. The snow scheme's layers can grow and collapse, with the number of layers and a layer's thickness being dependent on snow depth. Any solid precipitation (i.e., snowfall) is automatically integrated with surface snow or soil layers, and any rainfall is incorporated after fluxes and temperatures are updated [32].

Snow cover fraction affects surface albedo, net shortwave radiation, and almost all subsequent model energy and temperature calculations. Despite being such an important model component, it is parameterized in CLM2 and other LSMs as a predicted variable, dependent mainly on total snow depth (m). Therefore, any knowledge of snow cover from the previous timestep is only retained via the depth of the snowpack. CLM2's SCF calculation employs a single, nonlinear formula that is a function of both the snow depth state and momentum of roughness parameter (m) for soil, which is set to a default value of 0.01 m. The snow fraction value is then used in the overall direct and diffuse beam ground albedo calculations and subsequent energy and moisture flux predictions. Comparison studies have shown CLM2 to perform well in different regions and settings, especially for snow [31,37,38]. Subsequent versions of CLM (3.5 and greater) have incorporated additional updates to the snow physics with varying improvement (e.g., [20,39]) and will be made available in future releases of LIS.

The meteorological forcing dataset used to drive the CLM2 simulations is from the North American Land Data Assimilation System (NLDAS) [40]. The University of Maryland (UMD) land classification dataset, derived from the Terra MODIS (collection 4) UMD land cover classification product is used [41]. The CLM2's plant function type classification scheme and parameters were mapped to the UMD classification. For the elevation map, the 1 km NED elevation dataset is used. For further details of the CLM2 model set up and its use for this study, please see Arsenault et al. [15].

3.3. Data Assimilation Method and Experiment Setup

To assimilate the MODIS snow cover fraction estimates into CLM2, the 1-D Ensemble Kalman Filter (EnKF) method is used in this study, as adapted from Reichle et al. [42,43], which is also part of the LIS DA framework [44]. The SWE and snow depth states are perturbed, along with forcing fields that more effectively drive snow dynamics (e.g., precipitation, shortwave radiation), to generate the ensemble of model forecast conditions with 12 members. The DA methods and parameters used in this study are further outlined and described in De Lannoy et al. [14] and Arsenault et al. [15]. The MODIS SCF observations are also perturbed, as done with the EnKF method. The Terra MODIS SCF product is assimilated into CLM2 close to its local overpass time near 10:00 am local overpass time. Additional details of the snow cover assimilation, observation operators, and observation errors are further described in Sections 4 and 5.

An open-loop (OL) simulation was generated, which is a baseline simulation where no observations are assimilated, and is used as the benchmark experiment to compare the assimilated runs against. The CLM2 OL and DA experiments include a spin-up time period from 30 September 1999, to 30 September 2003. For the OL and DA experiments, a slightly smaller number of stations were used, since CLM2 was found to build "glacier-like" points (i.e., snowpack does not completely melt off during the summer months), due to (1) a cold-bias in the NLDAS temperature field, especially when brought to finer elevation scales (e.g., 1 KM); and (2) an absence of blowing snow represented in the model, which can be a major snow removal process near mountain terrain peaks [45]. Eliminating these "glacier" type points from the DA and analysis restricts the evaluation of the new SDC-based observation operators to the more normally varying snow accumulation and melt grid points. The reduction in stations resulted in a final 40 stations for WA and 78 for CO, versus the original 56 and 98, respectively, used in deriving the observation-based SDCs. These final stations and corresponding gridcells are highlighted with open grey circles in Figure 1, and they are used for the DA experiments and statistical results, shown in Section 5.

4. Methods

4.1. Deriving the Snow Depletion Curves from Snow Observations

To help capture the range of observed SWE and SCF relationships, a histogram analysis is introduced that allows for reducing the dispersion of data and retaining most measurements within the analysis. A binned scatterplot method is used where SNOTEL SWE observations (along the x-axis) are grouped in to bins of equal length (e.g., 40 mm of SWE) and the corresponding MODIS SCF observations are averaged within the given bin. This approach allows for highlighting relationships that may otherwise be masked in a normal scatter plot (e.g., [46,47]). For each bin, a boxplot can be produced to reflect the spread in MODIS SCF values for that given bin and be represented by that bin's statistics (mean, median, minimum/maximum, etc.). Figure 2 presents the full scatter across the different SNOTEL SWE and MODIS SCF observation ranges and the SCF averages per 40 mm SWE bin for the WA and CO domains. For this study, the average SCF (arithmetic mean) of each SWE bin is used.

For the period of interest, WYs 2000–2010, two years were selected, WY2004 and 2005, for the validation period, since they represent a range of snow conditions for the two domains. For both CO and WA, WY2004 experienced slightly less than average annual SWE. However, for WY2005, CO experienced above normal SWE and WA had very low SWE amounts. The years used for the binning and SDC equations include WYs 2000–2003 and 2006–2010, which are considered the "training" dataset. The WYs 2004–2005 are only included for the main DA experiments and validation period (see Section 5).

Figure 2 shows the range of MODIS SCF values (y-axis) in relation to the SNOTEL SWE points (x-axis) as scatter points. The large range of observed SCF and SWE values is noted here as a representative of the different snowpack and cover conditions that can occur in mountainous regions, e.g., low SWE amounts with high SCF during the accumulation phase. MODIS SCF values of 0% are excluded, and zero-valued SWE values are not included in the 40 mm bins. Some bins are removed from the curves if less than five MODIS SCF measurements were in a bin or were considered extreme outliers, resulting in a total of four bins removed for WA region and one for the CO region. The WA region scatterplot points (red) reveal much higher SNOTEL SWE values (greater than 2500 mm) than the CO SWE points. Since the points are scattered and weighted in a logarithmic fashion, logarithmic functions were fitted to the points to estimate relationships between the values. The WA logarithmic-regression line (red) indicates less change in SCF as SWE increases than the CO regressed line (green). Regression equations and coefficient of determination, R^2, values are highlighted in the lower right-hand corner boxes, where the red (green) box corresponds to WA (CO). The logarithmic equations and R^2 values shown here were generated within a Microsoft Excel Spreadsheet analysis. For the binned average values, logarithmic functions are linearly regressed onto them, producing high R^2 values: 0.822 and 0.931 for WA and CO, respectively. With binning the SWE observations and averaging the corresponding SCF values within each bin, much of the high data variability and spread is reduced with the binned averages, thus providing a greater "signal", as reflected with the R^2 values.

To show the robustness of this approach, other histogram tests were applied to the data to see if the relationships between the SWE bin and SCF average would change with different bin sizes or input data. First, an equal number of SNOTEL SWE observations were grouped per bin (1000 per bin) with SCF averaged across each new SWE bin. Another test was also applied where the SNOTEL observation sites were separated into two separate groups, thus two groups of 23 for WA (56 total sites) and 49 for CO (98 total) were set. These tests provided support for the use of the binned scatterplot approach in deriving the SDC-type relationships between the SNOTEL SWE and Terra MODIS SCF observations. For more details of these results, please see Section S1 of the Supplementary Material.

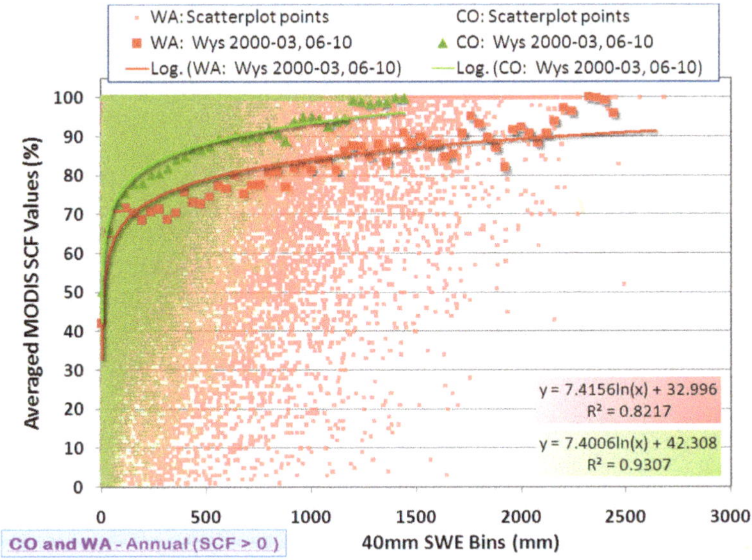

Figure 2. The binned scatterplot bin averages for Washington (WA) (large red squares) and Colarado (CO) (larger green triangles) imposed over all scatter plot points that are used in generating the bins. Logarithmic functions are fitted here to the binned snow cover fraction (SCF) averages with the red (green) line corresponding to the equation and R^2 value in the lower right-hand side red (green) box for WA (CO).

4.2. Applying the New Snow Depletion Curves as Observation Operators

In this section, the focus is on incorporating the logarithmic fitted functions, of the SWE-bin averaged SCF values, as new observation operators within the EnKF method. The observation operator used in the EnKF and other DA approaches, is typically represented with the symbol, **H** or **h**, which is used here as a vector function to transform the model SWE or snow depth states into the predicted SCF estimates, equivalent to the SCF observations. These two new annual-based, SDC-type observation operators for both the WA and CO regions are considered observation-based and referred to hereinafter as *obs*-**h** functions. We compare these new observation-based operators with the default, CLM2 model-based snow depletion curve, referred to hereinafter as the *model*-**h** observation operator. The 40 mm SWE binned SCF averages are plotted against predicted SCF values estimated with the CLM2 function, which is used as the default model-based SDC or observation-operator. The default CLM2 SDC function requires snow depth as the main input to calculate SCF. Thus, to compare it and its shape against the new observation-based SDC logarithmic functions on the same plot, we convert the 40 mm SWE bin values to snow depth, which requires a snow density estimate. Figure 3 compares these different SDC observation operators over the range of the 40 mm SWE-bin values, used as inputs to the model curve. For CLM2, three different bulk snow densities (100, 250, 400 kg m^{-3}) were used to derive snow depth values from the SWE bin inputs to show the impact on the predicted SCF values in relation to the observation-based functions.

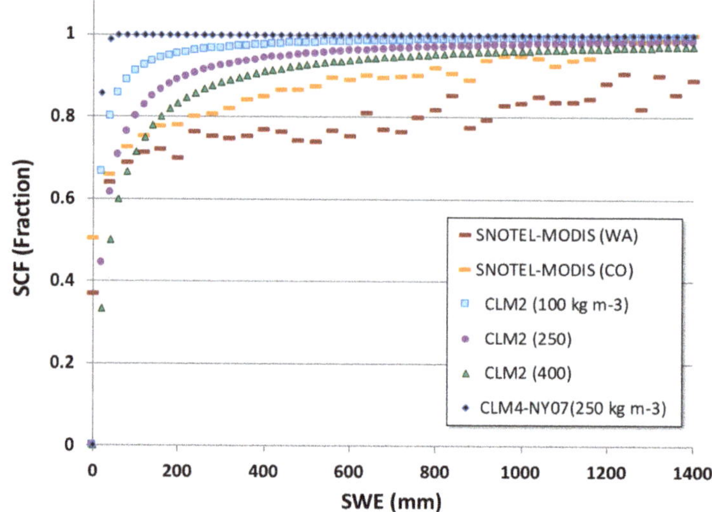

Figure 3. Comparison of the different SDC-observation operators (**h**), including the Community Land Model-based SDCs (CLM2 and CLM4-Niu and Yang (2007; NY07), and the observation-based observation operators (*obs*-**h**) for WA (maroon dashes) and CO (orange dashes). The 40 mm snow water equivalent (SWE) values are used in the CLM curves (*model*-**h**) to derive SCF estimates for this comparison. For the CLM2 curve, three different snow densities are used to show snow-depth to SWE (100, 250, 400 kg m^{-3}) ranges in SCF estimates, and the NY07 curve for just the 250 kg m^{-3} snow density value is used (navy blue diamonds).

In this comparison, the CLM2 range of density-dependent curves tend to have a greater rate of change in SCF for the first 400 mm of SWE in relation to the range of observation-based points (for both WA and CO). These differences translate into how model predicted SWE is converted into predicted SCF for the EnKF updates. In addition to the default CLM2 SDC function, the CLM version 4 (CLM4) SDC function, developed by Niu and Yang (2007; referred here as NY07), is included in Figure 3, using its default settings that have been used in some global climate model simulations [17] and snow cover DA studies [13]. For the CLM4 curve, only the bulk snow density of 250 kg m^{-3} was used, and shown in Figure 3; however, observation-based snow densities have been estimated to be between 175 and 320 kg m^{-3} at coarser scales (at $1.0° \times 1.0°$) by Niu and Yang [20]. For their data assimilation experiments, Su et al. [13] tuned the melt factor shape parameter and set the minimum snow density to 100 kg m^{-3}. The main difference in the CLM4 equation and that used in CLM2 is that the CLM4 hyperbolic tangent function reaches a value of 1, whereas the CLM2 function only approaches 1. Other studies have used the NY07 SDC type formulation for snow cover DA experiments, and either tuned the parameters or used the default model formulation [16].

The CLM4 NY07 SDC function would predict very high SCF values, even overpredict, for higher resolutions applications, such as used in this study with a 0.01° grid size. The CLM4 SDC was originally tuned to coarser scale SCF and snow depth observations (e.g., 1.0° grid size), so the higher predicted SCF values, shown in Figure 3, would not be as applicable to our high-resolution snow cover data assimilation study. The observation-based SDC logarithmic functions do suggest lower predicted SCF, with WA averages barely extending above 90% fractional snow coverage. Though it is argued here that the observation-based SDCs may reflect more realistic SCF estimates, these curves do not take in to account specific density changes, nor underestimation that can occur with the presence of dense forests and highly variable terrain, both major issues that impact MODIS SCF detection. The noticeable differences between the WA and CO binned SCF averages is most likely a result of such detection issues in WA, where there are especially much thicker and darker forest canopies than found in CO.

Next, we apply these different observation operators with the EnKF method and added new options in LIS to handle the two new *obs*-**h** functions for WA and CO. These two functions are considered representative of annually based climatologies and are therefore considered static in nature, encompassing a range of snow conditions, similar to the CLM2 *model*-**h** curves. In Figure 2, the two logarithmic functions shown were derived using a minimum of five observations per histogram bin. To impose a slightly stricter constraint in generating these equations, a minimum of ten observations per SWE bin was applied in generating the final SDC-type observation operator equations used in the data assimilation step. The final two annual observation-based *obs*-**h** logarithmic equations are:

$$WA\text{: } SCF_i^f = 5.7928 * \ln(SWE_i^f) + 43.0622, \tag{1}$$

$$CO\text{: } SCF_i^f = 6.8359 * \ln(SWE_i^f) + 45.8553, \tag{2}$$

where f denotes the model forecast and i denotes the ensemble member. Though the slope intercepts of these two logarithmic Equations (1) and (2), and what is shown in Figures 2 and 3, occur around 40 to 50% SCF when SWE is shown to be at the 0 value. However, when the model predicted SWE is at 0., the predicted SCF is also set to 0. Thus, when the predicted SWE goes above 0, Equation (1) or Equation (2), depending on the region, is applied as the observation operator in the DA innovation step, when comparing with the MODIS SCF estimate.

4.3. Generating Temporal and Physiographic Varying Snow Depletion Curves

Physiographic conditions are known to be a major control on larger spatial patterns of snow cover distribution, especially topography [2,3,48,49]. Also, snow cover patterns vary greatly throughout the year with extensive (lower) snow cover and low (high) snowpack depths in fall (spring), affected by snow accumulation (snowmelt) processes. As similarly derived for the annually representative observation-based SDC equations (described in the two previous sections), additional *obs*-**h** SDC relationships are derived and evaluated for different temporal and physiographic conditions, e.g., elevation bands and vegetation type, for the two regions, Washington and Colorado. The observation-based operators derived in the previous section will be referred to hereinafter as the *annual* based *obs*-**h** operators, to distinguish them from the ones that are derived here for the varying conditions.

Similar to the description in Section 4.1, the binned scatterplot approach is applied, but for varying elevation bands, vegetation types, and calendar months for both WA and CO. The SNOTEL SWE and MODIS SCF measurements are used in the same manner but binned for specific cases. The scatter-bins and logarithmic functions are generated for each case (see Supplementary Material for more information). For time of year, an *obs*-**h** curve is derived for each month, and separately for each region. The summer months were found in different calculations to average near zero, so no observed curves were fit for this season, and the CLM2 *model*-**h** curve is used instead when a MODIS SCF observation is present. For vegetation type, only the vegetation types associated with the MODIS-based UMD landcover map and collocated with a SNOTEL station (related to the equivalent gridcell location) are included. For WA and CO, five overlapping UMD classes resulted from the group of SNOTEL sites, so bins and curves were derived for those types. These five landcover types include evergreen needleleaf, mixed forest, woodland, open shrub and grassland. For CO, there was also one deciduous broadleaf case found collocated with a SNOTEL site. This was not included in the final DA experiment and the mixed forest SDC was applied at this location.

For elevation bands, elevation ranges were based on SNOTEL elevation values with near average mid-points of elevation taken as a cutoff value to distinguish higher versus lower elevation bands, as similarly done in Andreadis and Lettenmaier [10] and Su et al. [13]. For WA, the average midpoint elevation is about 1500 m and used to set the lower elevation range as 1000–1500 m and higher elevation range as 1501–2000 m. For CO, the two elevation band ranges are 2500–3000 m for the lower and 3001–3600 m for the upper band. These two bands represent the range of elevation conditions involving the SNOTEL sites, even though higher elevations extend above the highest SNOTEL station. Applying the scatterbin plot approach to each group of elevation points, two sets of logarithmic curves are generated like before for each region, one reflecting a lower and the other a higher elevation band. Table 1 presents an overview of the different model experiments and naming conventions used.

Table 1. A summary table of the different baseline and data assimilation-based experiments and the naming convention used for each experiment. Same experiments were conducted for both WA and CO domains over the water years, 2004 and 2005.

Experiment Name	Description
Open-loop (OL)	The baseline ensemble model experiment with no snow cover assimilation performed.
EnKF: *Model*-h	Snow cover assimilation using EnKF and the default model-based SDC as the observation operator.
EnKF: Annual *obs*-h	Snow cover assimilation using EnKF and the annual observation-based SDC as the observation operator.
EnKF: Month-based *obs*-h	Snow cover assimilation using EnKF and the monthly-varying, observation-based SDCs as the observation operator.
EnKF: Vegetation-based *obs*-h	Snow cover assimilation using EnKF and the vegetation-type, observation-based SDCs as the observation operator.
EnKF: Elevation band-based *obs*-h	Snow cover assimilation using EnKF and the elevation-band, observation-based SDCs as the observation operator.

Additional details, figures, and formulas for the newly fitted curves for the different conditions and regions are provided in the Supplementary Material, Section S2, that accompany this paper. Please refer to that published material for how the vegetation type, elevation band, and 12-month binned and fitted SDC functions were derived.

4.4. Estimating the Observational Standard Errors

To estimate the total observational standard error, σ_{total}, associated with each of these SDC-based observation operators, a method is followed that is similar to what is outlined in Arsenault et al. (2013). Using SNOTEL SWE and MODIS SCF measurements for WYs 2000–2003 and WYs 2006–2010, the new estimated σ_{total} values for the *obs*-h are: 33.04% for WA, and 28.67% for CO. For the *model*-h estimated σ_{total} values, they are 37.60% for WA, and 27.3% for CO. Previous studies have used a static 10% observational standard error [10,13,16], while others used a more dynamic or calibrated set of observational errors when assimilating MODIS snow cover [14,17]. It is hypothesized that by using the MODIS SCF observations themselves to derive new SDC formulas for the EnKF approach, should bring the model SCF predictions in line with the SCF observations. In a way, this is considered an observation bias reduction approach for the filter, by using observation operators that more closely match the observational averages.

For the different temporal and physiographic type observation-based operators, corresponding observation standard errors, σ_{total}, were also estimated by using these new logarithmic curves along with the available SNOTEL SWE and MODIS SCF observations for the period, WYs 2000–2003 and 2006–2010, which is outside of the data assimilation experiment window. The new σ_{total} values are calculated for the different monthly, vegetation and elevation conditions and for each region by inputting the SNOTEL SWE values in to the new logarithmic functions and then compared to the MODIS SCF estimates. This is again to capture both the errors in the SCF detection rates and newly derived observation operator functions. As shown in Figure 4, the results for both CO (blue bars on the right) and WA (red bars on the left) are compared side-by-side to examine the regional differences in the estimated observation errors. CO-associated errors are lower overall than WA, especially during the winter months. When these observational errors are lower, again more dependence on the observations is given, and thus through the EnKF updates, the model will draw closer to the observations. Stronger responses to the SCF observations are then expected in CO than in WA region for the DA results.

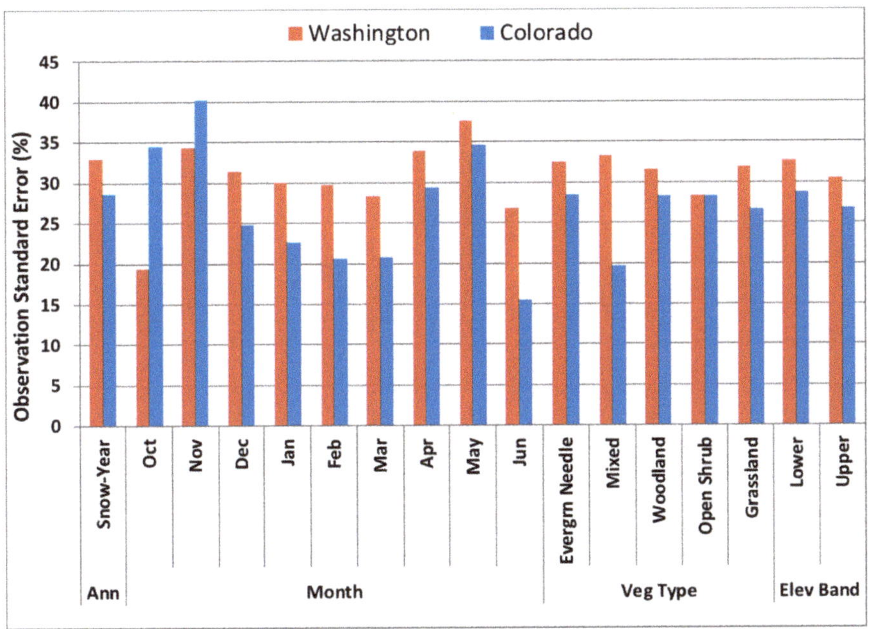

Figure 4. A bar plot of the varying observation standard error values, σ_{total}, (in % of SCF) for each given set of derived observation-based observation operators, *obs*-**h**, including annual-based, monthly, vegetation type and elevation band. Blue (red) bars indicate observation errors associated with CO (WA).

5. Results

5.1. Applying the SDC Observation Operators to the Data Assimilation Experiments

Using observations outside of the assimilation experiment period to derive the SDC formulas should test the new observation operators' ability to improve the snow state analysis. Here, the two *obs*-**h** based logarithmic functions, derived for WA and CO, are applied with the EnKF algorithm and compared with the *model*-**h** based EnKF experiments. As an initial comparison between these two EnKF experiments, Figures 5 and 6 provide timeseries of spatially averaged SWE and RMSE of SWE values (both in mm), respectively, across the SNOTEL validation points for the two regions and two Water Years (2004, 2005). For all four cases, the *obs*-**h** EnKF experiment produces higher averaged SWE conditions over the *model*-**h** EnKF run, especially for CO, bringing it in closer agreement with the SNOTEL spatially averaged SWE (Figure 5). The patterns of averaged accumulation and melt are similar to that of averaged SNOTEL SWE, just the magnitudes are lower for both the *model*-**h** and *obs*-**h** experiments, relative to the open-loop (OL) run. In addition, greater reduction in spatially averaged RMSE values corresponds to the *obs*-**h** based experiment (Figure 6), with most of the reduction in RMSE shown for CO.

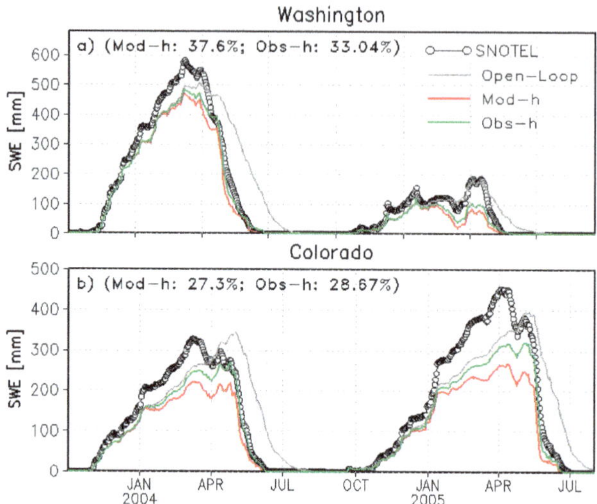

Figure 5. Spatially averaged SWE comparison is made between SNOTEL observations (open circles), the OL run (grey line), *model*-**h** EnKF experiment (red line), and *obs*-**h** EnKF (green line) experiments for both (**a**) WA and (**b**) CO regions. Values given within the parentheses indicate the σ_{total} values for the experiment.

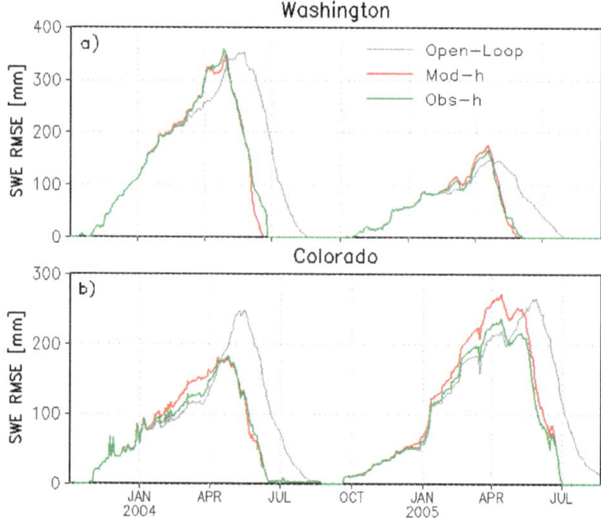

Figure 6. Spatially averaged RMSE SWE values (in mm) for both (**a**) WA and (**b**) CO for the OL run (grey line), *model*-**h** EnKF (red line) and *obs*-**h** EnKF (green line) experiments.

5.2. Annual- vs. Varying Conditions based Observation Operators

Previous snowcover DA studies have accounted for physiographic and seasonal-varying snow differences in their observation operators, however, they did not explore their overall impact on their EnKF experiments (e.g., [13]). In this section, the previously derived SDC logarithmic functions, representing different physiographic and temporal conditions, are explored as observation-based observation operators in a suite of EnKF experiments. Each group of physiographic- and temporal-varying conditions represents only a few categories (e.g., five UMD vegetation classes of thirteen total). The SDC function parameters, slope and intercept, and observational errors are mapped to other classes or bands similar in nature. For example, each vegetation class, like evergreen needleleaf forest, has its own set of function parameters and errors to better represent each set of separate conditions (e.g., snow conditions differ between leafless deciduous and needled conifer-based trees). Similarly, for elevation bands, elevations lower than the lower band limit used (e.g., an 800 m elevation pixel below the 1000 m lower limit for WA) use the function parameters and errors for that lower elevation band. Again, this might not truly reflect low elevation locations in mountainous areas, especially where MODIS SCA pixels are not known to detect snow cover presence as well (e.g., [50]).

By investigating the different conditions, e.g., vegetation types separate from elevation bands, we can see what individual impacts they may have versus all lumped together. For each region, three additional EnKF experiments were generated that incorporate the monthly, vegetation class, and elevation band type SDC functions, separately. Thus, for the EnKF experiment employing the vegetation class function parameters, the EnKF analysis updates will reflect the effects of those varying vegetation SDCs only. The same applies for the different months and elevation bands. To distinguish the different EnKF experiment types, they are simply named by their characteristic and that they are *obs*-**h** function types: Monthly, vegetation type (veg), and elevation (elev) *obs*-**h**, as highlighted in Table 1. The three different experiments are run for WYs 2004–2005, and compared with the *model*-**h** based EnKF experiments.

The first set of results are summarized in Table 2, which include overall means, standard deviation, RMSE, and correlation coefficient values between the simulation SWE results and SNOTEL SWE observations. These results are based on the reduced number of SNOTEL locations, 40 stations for WA and 78 for CO, since CLM2 was found to produce "glacier-like" points (i.e., lack of snowmelt in summer months) at the other 16 WA and 20 CO station gridcell locations. Table 2 shows lower RMSE values for many of the *obs*-**h** EnKF experiments, especially for CO in both years and less of an impact for WA, like WY2004. The lower RMSE values for CO's WY2005, noted as an anomalously positive snow year (from our eleven observation years), may be related to the curve parameters, especially for the vegetation type *obs*-**h** EnKF experiment, which would be dominated by lower SCF average values for CO. With it being a higher snow year than the 11-year average (2000–2010), higher MODIS SCF values would then "add" more snow to the model SWE analysis, thus resulting in higher SWE values and closer agreement to the observations.

To show how the RMSE values of each DA experiment relate to the open-loop run, Figure 7 shows each experiment's SWE-based RMSE values normalized with respect to those of the open-loop based on [51]. A normalized RMSE value less than one indicates the experiment may perform better than its control run, which is the open-loop in this case [51]. All experiments perform better than the open-loop runs, except for WA's WY2005, which was a major drought year and difficult to capture with the ensembles and observations. Also, the ± 1 standard deviation values (inner yellow error bars) in Figure 7 show smaller ranges for many of the *obs*-**h** experiments, especially the vegetation *obs*-**h** case for CO.

To see how this performance translates into EnKF metrics, the normalized innovation means and standard deviations for each experiment and water year are provided in Table 3. The means tend slightly closer to 0 for the *obs*-**h** than the *model*-h based experiments, indicating differences between the individual MODIS SCF values and model predicted SCF values were overall smaller. This may suggest that the *obs*-**h** curves predicted SCF values that were in slightly better agreement with the MODIS observations. The exception again was WY2005 for WA where there were greater differences, indicated by much higher negative mean values. This may relate to the shape of the curves and thus predicting higher SCF forecast values than the MODIS estimates. For the normalized innovation standard deviations, the values increase to near 1 with almost all the *obs*-**h** experiments, which may indicate that the model errors and observational errors may be more consistent in the filter (e.g., [52]) than those associated with the *model*-h EnKF experiments. This could possibly indicate better agreement in MODIS SCF values with the model predicted SCF produced by the *obs*-**h** curves.

Table 2. Comparison of summary SWE statistics for the period, October-June, between observation operator experiments. Bold numbers indicate experiments with reduced RMSE values relative to the open-loop and *model*-**h** EnKF experiments. SD refers to standard deviation.

	\multicolumn{8}{c}{**Observation Operator Comparison Statistics**}							
	Average	**SD**	**RMSE**	**Correl**	**Average**	**SD**	**RMSE**	**Correl**
	(mm)	**(mm)**	**(mm)**	**(-)**	**(mm)**	**(mm)**	**(mm)**	**(-)**
WA (*40 Sites*)			**2004**				**2005**	
Observations	221.85	216.19	-	-	61.19	66.56	-	-
Open-loop	247.01	188.17	168.22	0.79	64.30	61.14	65.09	0.74
EnKF: *Model*-h	183.54	184.21	138.54	0.87	34.62	40.38	60.29	0.75
EnKF: Annual *obs*-h	194.09	191.40	140.13	0.87	**40.11**	**45.38**	**57.46**	0.75
EnKF: Month *obs*-h	191.09	191.31	145.33	0.85	**37.27**	**42.89**	**58.29**	0.74
EnKF: Veg. *obs*-h	195.79	192.18	140.28	0.87	**40.60**	**45.77**	**57.31**	0.75
EnKF: Elev. *obs*-h	199.91	193.46	142.41	0.87	**40.80**	**46.07**	**57.83**	0.75
CO (*78 Sites*)			**2004**				**2005**	
Observations	132.74	127.26	-	-	187.20	168.07	-	-
Open-loop	143.75	115.63	107.00	0.72	180.14	132.52	127.97	0.79
EnKF: *Model*-h	94.01	92.34	81.57	0.89	119.57	109.58	116.57	0.89
EnKF: Annual *obs*-h	**108.00**	**105.82**	**79.24**	**0.88**	139.72	127.99	103.69	0.91
EnKF: Month *obs*-h	111.15	109.82	86.21	0.86	145.82	134.43	109.96	0.89
EnKF: Veg. *obs*-h	**114.26**	**111.18**	**76.91**	**0.88**	147.93	134.33	99.51	0.91
EnKF: Elev. *obs*-h	112.60	109.71	81.74	0.87	144.87	132.64	108.19	0.90

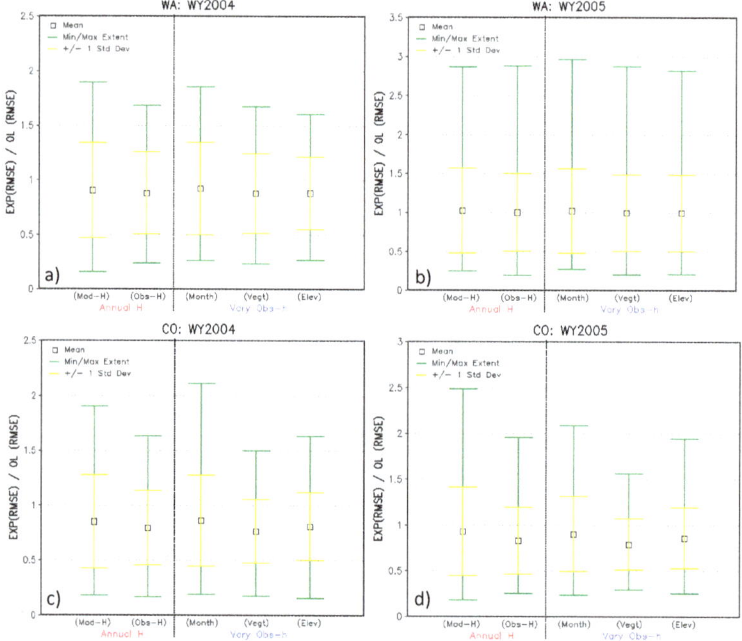

Figure 7. Comparison of each experiment's SWE RMSE values normalized by the open-loop (OL) RMSE values for the five experiments: *Model*-**h** EnKF run, and the annual, monthly, vegetation, and elevation varying *obs*-**h** EnKF experiments for WA (**a,b**) and CO (**c,d**) for WY2004 (left) and WY2005 (right). Black open squares indicate the statistical mean, yellow error bar lines indicate ±1 stdev unit from mean, and green error bar lines indicate the maximum and minimum extents of the normalized statistic. Values below 1 suggest improvement over the OL run.

Table 3. Table of normalized innovation values, including the mean and standard deviation (SD) for the different observation operator, *h*, experiments.

	Normalized Innovation Analysis							
	WA				CO			
	2004		2005		2004		2005	
Experiment	Mean	SD	Mean	SD	Mean	SD	Mean	SD
CLM2 *mod*-**h**	−0.43	0.89	−0.39	0.81	−0.43	0.92	−0.37	0.87
Annual *obs*-**h**	−0.41	0.99	−0.57	0.95	−0.42	0.98	−0.34	0.94
Monthly *obs*-**h**	−0.41	0.97	−0.59	0.98	−0.38	1.02	−0.31	0.93
Veg. *obs*-**h**	−0.41	1.02	−0.58	0.95	−0.39	1.01	−0.30	0.97
Elev. *obs*-**h**	−0.41	1.00	−0.57	0.95	−0.41	1.02	−0.32	0.99

6. Discussion

In this study, we have explored how different observation-based snow depletion curve (SDC) characteristics can be derived and used as observation-based observation operators in EnKF assimilation updates when assimilating MODIS, or other satellite-based, snow cover fraction estimates. Different SDC logarithmic functions, representing different physiographic and temporal conditions, were also explored as observation-based observation operators in a full suite of EnKF experiments. Using logarithmic functions with a y-axis intercept value, not set to 0, means that MODIS SCF values may not be assimilated below that value (e.g., 40% SCF). In a way, this acts as a cutoff for lower MODIS SCF values (e.g., less than 30%), which could contain contaminated SCF values, due to unresolved patchiness in a pixel and NDSI algorithm pixel discrimination issues [53]. Also, the fact that many of the logarithmic curves never fully reach 100% SCF could allow MODIS SCF pixels that are near 100% to have an impact on the snow analysis, when the model SCF forecasts are much lower than 100%. This situation can work well when the LSM consistently underestimates SWE, relative to the in situ SWE observations. Therefore, for LSMs with a low SWE bias or precipitation input bias (e.g., an undercatch issue), this SDC approach can partly compensate for the low bias, especially when assimilating snow cover in higher mountain catchments locations. However, if an LSM or snow model has a high SWE bias compared to the observations, then this observed type SDC could contribute to overestimating and adding too much snow to the model.

In other snow cover data assimilation studies, most curves are designed to reach 100% SCF, even for coarser grid scale experiments [13]. If model SWE forecast conditions remain high enough that the predicted SCF ensemble forecasts remain at 100%, with little ensemble spread in SCF, then there can be very little to no impact on the SWE analysis if the observed SCF is much less than 100%, e.g., partial coverage for the gridcell [14]. Another minor downside to having an SDC-type observation operator reach 100% snow cover, like the case for many LSM curves, the EnKF-based model SCF forecast ensemble can become underestimated, if perturbations force the members to fall below 100%, even if both the MODIS and model predicted SCF were originally at 100%. This can also occur when perturbing the MODIS SCF observations. When the observed SCF is at 100%, this can also reduce that value when perturbed, leading to underestimated SWE analysis [14]. Rules can be applied in the EnKF method as to how the ensemble members, e.g., within the observation-perturbed ensemble, get distributed or updated near the 100% SCF point. This could include a set of rules for reducing the SWE analysis by a certain fraction when there is a partial amount of observed SCF or by modifying the observation error covariance, σ_{total}, that controls the ensemble or uncertainty spread [based on 14]. One slight advantage to the SNOTEL-MODIS observation-based observation operators, *obs*-**h**, is that they reflect an average SCF percentage of what the satellite detects for a given range of conditions. Thus, if the predicted SCF is 80% and is the upper threshold in the SDC function, then 20% of the area could be considered exposed vegetation cover or average SCF conditions, for say, lower elevation regions.

Based on the filter statistics shown in Tables 2 and 3, there still remains some large differences between the magnitudes of the model predicted SCF and the MODIS SCF observations. One approach to address these innovation biases would be to scale the MODIS SCF values to reflect higher ones, bringing them into closer agreement with the predicted SCF values [54]. Another option would be to place additional constraints on the observation operators or SCF observations themselves, or to further tune the observations or the curves towards the other, so that the filter statistics reflect less bias between observations and the model. In this study, by deriving the observation-based observation operators with the MODIS SCF observations, it helps to bring the predicted SCF values closer in agreement to the observations. With almost all data assimilation approaches, tuning may be the unavoidable course to be taken, so the final analysis may reflect the best that both the model and observations have to offer.

7. Conclusions

In this study, we presented a new approach to estimating snow depletion curves and their application for assimilating snow cover fraction observations, using an EnKF approach and a land surface model with a multi-layer snow physics scheme. We use observed SWE and snow cover fraction estimates to derive new SDCs, using a larger array of observations than previous studies, spanning two different mountainous regions in the U.S., in Washington and Colorado. We refer to these new SDC observation operators as "observation-based" and benchmarked their skill against the default, model-based SDC function. The SDC observation-based observation operators showed improvement over the default model-based SCF forecasts and snow state analysis. A secondary goal of this study was to apply this SDC-type approach to see how accounting for varying vegetation, elevation, and temporal conditions may better capture heterogeneous features related to the snowpack and snow cover patterns when assimilating snow cover observations. Vegetation-based curves showed improvement over the lumped annual observation-based SDC, especially for Colorado. Finally, the new SDC-based observation operators are used to derive the observational errors, which were used in the EnKF method's snow cover observation perturbations. Accounting for different errors related to the varying conditions provides different weighting of the snow cover observations against that of the model in the EnKF innovation and update steps.

These results along with other previous SCA DA studies do show promise in that there are positive impacts without much tuning, but additional testing with the observation operators, model, and observations is still needed [10,18]. In future work, more a priori information about the MODIS SCF predictions could be included to deriving and tuning functions, like the ones developed here for this DA application. For example, determining error information with regard to MODIS SCF 100% estimates in connection with the meteorological forcing fields, e.g., ability to estimate snowfall for same observed 100% SCF events or the role that temperature can play in MODIS SCA errors [55]. Also, more adaptive type SDC schemes, which could be more representative at a point, could be developed that adjust relative to changing conditions. Different approaches could be applied to optimize these observational based SDCs and improve the overall performance of the EnKF system.

Supplementary Materials: The following are available online at http://www.mdpi.com/2076-3263/8/12/484/s1, Figure S1: Presenting the binned scatterplots for the two mountain regions. Figure S2: Showing the robustness of the scatterbin histogram approach for deriving the SDCs. Figure S3: Monthly SDC relationships derived from binned scatterplot approach for the Washington state domain. Figure S4: Similar to Figure S3, but for Colorado state domain. Figure S5: SDC relationships derived for two different elevation bands for Washington domain. Figure S6: Similar to Figure S5, but for the Colorado domain. Figure S7: SDC relationships derived for different vegetation types from binned scatterplot approach for Washington. Figure S8: Similar to Figure S7, but for Colorado. Figure S9: Independent check on the snow water equivalent bins for different dominant vegetation types and in relation to elevation, for Washington domain. Figure S10: Similar to Figure S9, but for Colorado. Figure S11: Timeseries comparison of the experiments' spatially averaged predicted SCF (in %) for the different observation operators and SCF observations, presented for the melt season. Figure S12: Individual SNOTEL site comparisons between different SDC-type observation operators.

Author Contributions: Much of the work, including methodology, code development, validation and analysis, was performed by K.R.A., and overall supervision and funding acquisition was provided by P.R.H. The original draft preparation and writing was done by K.R.A., and review and editing was done by P.R.H.

Funding: This research was originally funded by grants given by the U.S. National Oceanic and Atmospheric Administration and the National Aeronautics and Space Administration.

Acknowledgments: We want to acknowledge the original support and computing resources provided by The Center for Ocean-Land-Atmosphere Studies (COLA) at George Mason University in Fairfax, Virginia, U.S.A.

Conflicts of Interest: The authors declare no conflict of interest.

References

1. Qu, X.; Hall, A. Assessing Snow Albedo Feedback in Simulated Climate Change. *J. Clim.* **2006**, *19*, 2617–2630. [CrossRef]
2. Fassnacht, S.R.; Sexstone, G.A.; Kashipazha, A.H.; Lopez-Moreno, J.I.; Jasinski, M.F.; Kampf, S.K.; Von Thaden, B.C. Deriving snow-cover depletion curves for different spatial scales from remote sensing and snow telemetry data. *Hydrol. Process.* **2016**, *30*, 1708–1717. [CrossRef]
3. Driscoll, J.M.; Hay, L.E.; Bock, A. Spatiotemporal Variability of Snow Depletion Curves Derived from SNODAS for the Conterminous United States, 2004–2013. *J. Am. Water Resour. Assoc.* **2017**, *53*, 655–666. [CrossRef]
4. Liston, G.E. Representing subgrid snow cover heterogeneities in regional and global models. *J. Clim.* **2004**, *17*, 1381–1397. [CrossRef]
5. Anderson, E.A. *National Weather Service River Forecast System: Snow Accumulation and Ablation Model*; NWS HYDRO-17; NOAA Technical Memorandum; NOAA: Silver Spring, MD, USA, 1973.
6. Donald, J.R.; Soulis, E.D.; Kouwen, N.; Pietroniro, A. A land cover-based snow cover representation for distributed hydrologic-models. *Water Resour. Res.* **1995**, *31*, 995–1009. [CrossRef]
7. Kolberg, S.A.; Gottschalk, L. Updating of snow depletion curve with remote sensing data. *Hydrol. Process.* **2006**, *20*, 2363–2380. [CrossRef]
8. Shamir, E.; Georgakakos, K.P. Estimating snow depletion curves for American River basins using distributed snow modeling. *J. Hydrol.* **2007**, *334*, 162–173. [CrossRef]
9. DeBeer, C.M.; Pomeroy, J.W. Influence of snowpack and melt energy heterogeneity on snow cover depletion and snowmelt runoff simulation in a cold mountain environment. *J. Hydrol.* **2017**, *553*, 199–213. [CrossRef]
10. Andreadis, K.M.; Lettenmaier, D.P. Assimilating remotely sensed snow observations into a macroscale hydrology model. *Adv. Water Resour.* **2006**, *29*, 872–886. [CrossRef]
11. Kolberg, S.; Rue, H.; Gottschalk, L. A Bayesian spatial assimilation scheme for snow coverage observations in a gridded snow model. *Hydrol. Earth Syst. Sci.* **2006**, *10*, 369–381. [CrossRef]
12. Zaitchik, B.; Rodell, M. Forward-looking assimilation of MODIS-derived snow-covered area into a land surface model. *J. Hydrometeor.* **2009**, *10*, 130–148. [CrossRef]
13. Su, H.; Yang, Z.-L.; Niu, G.-Y.; Dickinson, R.E. Enhancing the estimation of continental-scale snow water equivalent by assimilation MODIS snow cover with the ensemble Kalman filter. *J. Geophys. Res.* **2008**, *113*, D08120. [CrossRef]
14. De Lannoy, G.J.M.; Reichle, R.H.; Arsenault, K.R.; Houser, P.R.; Kumar, S.; Verhoest, N.E.C.; Pauwels, V.R.N. Multiscale assimilation of Advanced Microwave Scanning Radiometer-EOS snow water equivalent and Moderate Resolution Imaging Spectroradiometer snow cover fraction observations in northern Colorado. *Water Resour. Res.* **2012**, *48*, W01522. [CrossRef]
15. Arsenault, K.R.; Houser, P.R.; De Lannoy, G.J.M.; Dirmeyer, P.A. Impacts of snow cover fraction data assimilation on modeled energy and moisture budgets. *J. Geophys. Res. Atmos.* **2013**, *118*, 7489–7504. [CrossRef]
16. Zhang, Y.-F.; Hoar, T.J.; Yang, Z.-L.; Anderson, J.L.; Toure, A.M.; Rodell, M. Assimilation of MODIS snow cover through the Data Assimilation Research Testbed and the Community Land Model version 4. *J. Geophys. Res. Atmos.* **2014**, *119*, 7091–7103. [CrossRef]
17. Xu, J.; Zhang, F.; Shu, H.; Zhong, K. Improvement of the Snow Depth in the Common Land Model by Coupling a Two-Dimensional Deterministic Ensemble Model with a Variational Hybrid Snow Cover Fraction Data Assimilation Scheme and a New Observation Operator. *J. Hydrometeorol.* **2017**, *18*, 119–138. [CrossRef]

18. Clark, M.P.; Slater, A.G.; Barrett, A.P.; Hay, L.E.; McCabe, G.J.; Rajagopalan, B.; Leavesley, G.H. Assimilation of snow-covered area information into hydrologic and land-surface models. *Adv. Water Resour.* **2006**, *29*, 1209–1221. [CrossRef]

19. Thirel, G.; Salamon, P.; Burek, P.; Kalas, M. Assimilation of MODIS snow cover area data in a distributed hydrological model using the particle filter. *Remote Sens.* **2013**, *5*, 5825–5850. [CrossRef]

20. Niu, G.-Y.; Yang, Z.-L. An observation-based formulation of snow cover fraction and its evaluation over large North American river basins. *J. Geophys. Res.* **2007**, *112*, D21101. [CrossRef]

21. Salomonson, V.V.; Appel, I. Estimating fractional snow cover from MODIS using the normalized difference snow index. *Remote Sens. Environ.* **2004**, *89*, 351–360. [CrossRef]

22. Hall, D.K.; Riggs, G.A. Accuracy Assessment of the MODIS snow products. *Hydrol. Process.* **2007**, *21*, 1534–1547. [CrossRef]

23. Salomonson, V.V.; Appel, I. Development of the Aqua MODIS NDSI fractional snow cover algorithm and validation results. *IEEE Trans. Geosci. Remote Sens.* **2006**, *44*, 1747–1756. [CrossRef]

24. Hall, D.K.; Riggs, G.A. *MODIS/Terra Snow Cover Daily L3 Global 500 m Grid, Version 5*; NASA National Snow and ICE Data Center Distributed Active Archive Center: Boulder, CO, USA, 2011. [CrossRef]

25. Parajka, J.; Holko, L.; Kostka, Z.; Bloschl, G. MODIS snow cover mapping accuracy in a small mountain catchment—Comparison between open and forest sites. *Hydrol. Earth Syst. Sci.* **2012**, *16*, 2365–2377. [CrossRef]

26. Dwyer, J.; Schmidt, G. The MODIS Reprojection Tool. In *Earth Science Satellite Remote Sensing*; Qu, J.J., Gao, W., Kafatos, M., Murphy, R.E., Salomonson, V.V., Eds.; Springer: Berlin/Heidelberg, Germany, 2006.

27. National Climatic Data Center. *Data Documentation for Dataset 6430 (DSI-6430), USDA/NRCS SNOTEL—Daily Data*; Text Reports; NCDC: Asheville, NC, USA, 2002.

28. Pan, M.; Sheffield, J.; Wood, E.F.; Mitchell, K.E.; Houser, P.R.; Schaake, J.C.; Robock, A.; Lohmann, D.; Cosgrove, B.; Duan, Q.; et al. Snow process modeling in the North American Land Data Assimilation System (NLDAS): 2. Evaluation of model simulated snow water equivalent. *J. Geophys. Res.* **2003**, *108*. [CrossRef]

29. Serreze, M.C.; Clark, M.P.; Armstrong, R.L.; McGinnis, D.A.; Pulwarty, R.S. Characteristics of the western United States snowpack from snowpack telemetry (SNOTEL) data. *Water Resour. Res.* **1999**, *35*, 2145–2160. [CrossRef]

30. Bonan, G.B.; Oleson, K.W.; Vertenstein, M.; Levis, S.; Zeng, X.; Dai, Y.; Dickinson, R.E.; Yang, Z.-L. The land surface climatology of the Community Land Model coupled to the NCAR Community Climate Model. *J. Clim.* **2002**, *15*, 3123–3149. [CrossRef]

31. Dai, Y.; Zeng, X.; Dickinson, R.E.; Baker, I.; Bonan, G.B.; Bosilovich, M.G.; Denning, A.S.; Dirmeyer, P.A.; Houser, P.R.; Niu, G.; et al. The Common Land Model. *Bull. Am. Meteorol. Soc.* **2003**, *84*, 1013–1023. [CrossRef]

32. Oleson, K.W.; Dai, Y.; Bonan, G.; Bosilovich, M.; Dickinson, R.; Dirmeyer, P.; Hoffman, F.; Houser, P.; Levis, S.; Niu, G.-Y.; et al. *Technical Description of the Community Land Model (CLM)*; NCAR Technical Note; NCAR/TN-461+STR; Center for Atmospheric Research: Boulder, CO, USA, 2004.

33. Kumar, S.V.; Peters-Lidard, C.D.; Tian, Y.; Houser, P.R.; Geiger, J.; Olden, S.; Lighty, L.; Eastman, J.L.; Doty, B.; Dirmeyer, P.; et al. Land Information System—An Interoperable Framework for High Resolution Land Surface Modeling. *Environ. Model. Softw.* **2006**, *21*, 1402–1415. [CrossRef]

34. Anderson, E.A. *A Point Energy and Mass Balance Model of a Snow Cover*; NOAA Technical Report NWS 19; Office of Hydrology, National Weather Service: Silver Spring, MD, USA, 1976.

35. Jordan, R. *A One-Dimensional Temperature Model for a Snow Cover: Technical Documentation for SNTHERM.89*; Special Report 91-16; U.S. Army Cold Regions Research and Engineering Laboratory: Hanover, NH, USA, 1991.

36. Dai, Y.; Zeng, Q. A land surface model (IAP94) for climate studies. Part I: Formulation and validation in off-line experiments. *Adv. Atmos. Sci.* **1997**, *14*, 433–460.

37. Feng, X.; Sahoo, A.; Arsenault, K.; Houser, P.; Luo, Y.; Troy, T.J. The Impact of Snow Model Complexity at Three CLPX Sites. *J. Hydrometeorol.* **2008**, *9*, 1464–1481. [CrossRef]

38. Rutter, N.; Essery, R.; Pomeroy, J.; Altimir, N.; Andreadis, K.; Baker, I.; Barr, A.; Bartlett, P.; Boone, A.; Deng, H.; et al. Evaluation of forest snow process models (SnowMIP2). *J. Geophys. Res.* **2009**, *114*, D06111. [CrossRef]

39. Wang, A.; Zeng, X. Improving the treatment of vertical snow burial fraction over short vegetation in the NCAR CLM3. *Adv. Atmos. Sci.* **2009**, *26*, 877–886. [CrossRef]

40. Cosgrove, B.A.; Lohmann, D.; Mitchell, K.E.; Houser, P.R.; Wood, E.F.; Schaake, J.C.; Robock, A.; Marshall, C.; Sheffield, J.; Duan, Q.; et al. Real-time and retrospective forcing in the North American Land Data Assimilation System (NLDAS) project. *J. Geophys. Res.* **2003**, *108*, 8842. [CrossRef]

41. Friedl, M.A.; McIver, D.K.; Hodges, J.C.F.; Zhang, X.Y.; Muchoney, D.; Strahler, A.H.; Woodcock, C.E.; Gopal, S.; Schneider, A.; Cooper, A.; et al. Global land cover classification at 1km spatial resolution using a classification tree approach. *Remote Sens. Environ.* **2002**, *83*, 287–302. [CrossRef]

42. Reichle, R.H.; McLaughlin, D.B.; Entekhabi, D. Hydrologic data assimilation with the Ensemble Kalman filter. *Mon. Weather Rev.* **2002**, *130*, 103–114. [CrossRef]

43. Reichle, R.H.; Bosilovich, M.G.; Crow, W.T.; Koster, R.D.; Kumar, S.V.; Mahanama, S.P.P.; Zaitchik, B.F. Recent Advances in Land Data Assimilation at the NASA Global Modeling and Assimilation Office. In *Data Assimilation for Atmospheric, Oceanic and Hydrologic Applications*; Park, S.K., Xu, L., Eds.; Springer: New York, NY, USA, 2009; pp. 407–428. [CrossRef]

44. Kumar, S.V.; Reichle, R.H.; Peters-Lidard, C.D.; Koster, R.D.; Zhan, X.; Crow, W.T.; Eylander, J.B.; Houser, P.R. A land surface data assimilation framework using the land information system: Description and applications. *Adv. Water Resour.* **2008**, *31*, 1419–1432. [CrossRef]

45. Bowling, L.C.; Pomeroy, J.W.; Lettenmaier, D.P. Parameterization of blowing-snow sublimation in a macroscale hydrology model. *J. Hydrometeorol.* **2004**, *5*, 745–762. [CrossRef]

46. Jacko, J.A.; Stephanidis, C. (Eds.) *Human-Computer Interaction: Theory and Practice*; Lawrence Erlbaum and Associates: Mahwah, NJ, USA, 2003.

47. Seo, H.; Miller, A.J.; Roads, J.O. The Scripps Coupled Ocean-Atmosphere Regional (SCOAR) Model, with Applications in the Eastern Pacific Sector. *J. Clim.* **2007**, *20*, 381–402. [CrossRef]

48. Jost, G.; Weiler, M.; Gluns, D.R.; Alila, Y. The influence of forest and topography on snow accumulation and melt at the watershed-scale. *J. Hydrol.* **2007**, *347*, 101–115. [CrossRef]

49. Anderton, S.P.; White, S.M.; Alvera, B. Evaluation of spatial variability in snow water equivalent for a high mountain catchment. *Hydrol. Process.* **2004**, *18*, 435–453. [CrossRef]

50. Molotch, N.P.; Margulis, S.A. Estimating the distribution of snow water equivalent using remotely sensed snow cover data and a spatially distributed snowmelt model: A multi-resolution, multi-sensor comparison. *Adv. Water Res.* **2008**, *31*, 1503–1514. [CrossRef]

51. Crow, W.T.; Ryu, D. A new data assimilation approach for improving runoff prediction using remotely-sensed soil moisture retrievals. *Hydrol. Earth Syst. Sci.* **2009**, *13*, 1–16. [CrossRef]

52. Whitaker, J.S.; Hamill, T.M.; Wei, X.; Song, Y.; Toth, Z. Ensemble data assimilation with the NCEP Global Forecast System. *Mon. Weather Rev.* **2008**, *136*. [CrossRef]

53. Déry, S.J.; Salomonson, V.V.; Stieglitz, M.; Hall, D.K.; Appel, I. An approach to using snow areal depletion curves inferred from MODIS and its application to land surface modelling in Alaska. *Hydrol. Process.* **2005**, *19*, 2755–2774. [CrossRef]

54. Xu, L.; Dirmeyer, P. Snow-atmosphere coupling strength. Part I: Effect of model biases. *J. Hydrometeorol.* **2013**, *14*, 389–403. [CrossRef]

55. Dong, J.; Peters-Lidard, C. On the relationship between temperature and MODIS snow cover retrieval errors in the Western U.S. *IEEE J. Sel. Top. Appl. Earth Obs. Remote Sens.* **2010**, *3*, 132–140. [CrossRef]

 *geosciences*

Article

Advances in Snow Hydrology Using a Combined Approach of GNSS In Situ Stations, Hydrological Modelling and Earth Observation—A Case Study in Canada

Florian Appel [1,*], Franziska Koch [2,3], Anja Rösel [1], Philipp Klug [1], Patrick Henkel [4], Markus Lamm [4], Wolfram Mauser [1,3] and Heike Bach [1]

[1] VISTA Remote Sensing in Geosciences GmbH, Gabelsbergerstr. 51, D-80333 Munich, Gemany; roesel@vista-geo.de (A.R.); klug@vista-geo.de (P.K.); w.mauser@lmu.de (W.M.); bach@vista-geo.de (H.B.)

[2] Institute of Water Management, Hydrology and Hydraulic Engineering, University of Natural Resources and Life Sciences, 1190 Vienna, Austria; franziska.koch@boku.ac.at

[3] Ludwig-Maximilians-Universität, D-80539 Munich, Germany

[4] ANavS GmbH, D-80333 Munich, Germany; patrick.henkel@anavs.de (P.H.); markus.lamm@anavs.de (M.L.)

[*] Correspondence: appel@vista-geo.de; Tel.: +49-89-4521-614-15

Received: 8 January 2019 ; Accepted: 10 January 2019; Published: 15 January 2019

Abstract: The availability of in situ snow water equivalent (SWE), snowmelt and run-off measurements is still very limited especially in remote areas as the density of operational stations and field observations is often scarce and usually costly, labour-intense and/or risky. With remote sensing products, spatially distributed information on snow is potentially available, but often lacks the required spatial or temporal requirements for hydrological applications. For the assurance of a high spatial and temporal resolution, however, it is often necessary to combine several methods like Earth Observation (EO), modelling and in situ approaches. Such a combination was targeted within the business applications demonstration project SnowSense (2015–2018), co-funded by the European Space Agency (ESA), where we designed, developed and demonstrated an operational snow hydrological service. During the run-time of the project, the entire service was demonstrated for the island of Newfoundland, Canada. The SnowSense service, developed during the demonstration project, is based on three pillars, including (i) newly developed in situ snow monitoring stations based on signals of the Global Navigation Satellite System (GNSS); (ii) EO snow cover products on the snow cover extent and on information whether the snow is dry or wet; and (iii) an integrated physically based hydrological model. The key element of the service is the novel GNSS based in situ sensor, using two static low-cost antennas with one being mounted on the ground and the other one above the snow cover. This sensor setup enables retrieving the snow parameters SWE and liquid water content (LWC) in the snowpack in parallel, using GNSS carrier phase measurements and signal strength information. With the combined approach of the SnowSense service, it is possible to provide spatially distributed SWE to assess run-off and to provide relevant information for hydropower plant management in a high spatial and temporal resolution. This is particularly needed for so far non, or only sparsely equipped catchments in remote areas. We present the results and validation of (i) the GNSS in situ sensor setup for SWE and LWC measurements at the well-equipped study site Forêt Montmorency near Quebec, Canada and (ii) the entire combined in situ, EO and modelling SnowSense service resulting in assimilated SWE maps and run-off information for two different large catchments in Newfoundland, Canada.

Keywords: snow; SWE; LWC; run-off modelling; hydropower application; GNSS; EO

---

1. Introduction

The seasonal snow cover has an important role as hydrological storage for the Earth's fresh water resources. The amount of water stored in the snowpack as snow and ice is expressed as snow water equivalent (SWE) and is a key variable in water resources management, which is an essential component within the Earth's climate system [1]. The amount of water, which is released seasonally (or in events) as snowmelt in the rivers, as well as the timing of the water release, mainly in spring, is relevant for numerous hydrological applications, such as hydropower production, irrigation, and fresh water supply. In addition, knowledge about the snow situation is a concern of many safety related institutions and businesses, such as avalanche warning centres and (re-)insurance companies. The onset of snowmelt and its intensity are major drivers for flood forecasting, especially in mountainous areas, and are, besides the knowledge on the total amount of water stored as snow, a very valuable information for hydropower companies.

Regarding in situ snow measurements, until now, SWE is mainly measured manually by weighing a given volume of snow, which is cut out of the snowpack with tubes [2]. This approach is reliable but can provide only a snapshot in time and in space, and in addition, it is labour-intense and destructive. Automatic SWE measurements are mainly performed by weighing systems like snow pillows and snow scales [3]. These methods provide time series, but the instruments and their installation and operation are quite costly, and their results might be affected by ice-bridging effects of thermal fluxes resulting in potential over- or underestimations of SWE [4]. Alternative SWE in situ observation systems make, e.g., use of cosmic rays or neutron rays [5] or apply a passive gamma monitoring sensor (GMON) [6]. However, the reliability of those sensors depends strongly on the underlying surface conditions, the measurements are limited to a certain amount of SWE and thus are only applicable at certain locations [7].

As an alternative to the standard in situ methods, L-band Global Navigation Satellite (GNSS) signals can be used to derive snow cover properties. Different methods were developed within the last years. As an advantage, it is possible to apply these methods globally as the GNSS signals can be tracked almost any place on Earth, they are non-destructive and can be used even for large amounts of snow [8]. Snow height (HS), for example, can be determined by the reflection of GNSS signals on the snow–air interface [9]. Liquid water content (LWC) can be derived by GNSS signal strength attenuation through a snowpack of a given volume [10]. Henkel et al. [11] presented for the first time the possibility to derive SWE for dry snow conditions using two low-cost GNSS sensors for a carrier phase-based approach to detect signal changes within the snowpack. Steiner et al. [12] confirmed this by using a similar technique with geodetic sensors and applying different ambiguity resolution strategies and wideline combinations. Finally, we developed a novel approach combining GNSS signal attenuation and time delay by combining information on GNSS carrier phases and signal strengths. We accomplished deriving the three snow cover parameters SWE, LWC, and snow height in parallel, as recently demonstrated in Koch et al. [8].

Spatially distributed snow information such as the snow cover extent, information weather the snow is dry or wet, and dry snow SWE or snow height, can be derived from Earth Observation (EO) data, based on different remote sensing techniques using active or passive microwaves, or optical, infrared or thermal approaches. An overview is given, for example, by Hall [13] and Tedesco [14]. In recent years, especially the freely accessible Sentinel-1, -2 and -3 data are a useful source for determining the above-mentioned snow parameters, like snow extent, or wet snow [15]. Besides 'raw' satellite data, also different, often project-based, internet portals like GlobSnow, CryoLand, Google Earth Engine, or EUMETSAT H-SAF are providing already processed satellite-based snow parameter products. However, in general, all remote sensing products are often not available in high-temporal resolution or may lack the required spatial resolution. This is especially the case for optical images, e.g., from MODIS, which face potential cloud cover issues [16–18]. Active microwave products are often restricted due to foreshortening or layover effects, in particular in mountain regions and passive microwave products are very coarse regarding their spatial resolution. Recent approaches tend to apply more and more multi-sensor techniques to overcome some of these limits, which is, e.g., currently a big aim of the NASA SnowEx campaign [19]. Additionally, the combination of EO and hydrological model approaches helps to increase the temporal and spatial availability of snow or run-off information as e.g., presented by Cline et al. [20] and Immerzeel et al. [21].

In the current study, we present a comprehensive overview of a combined approach on using in situ measurements, EO, and hydrological modelling to derive continuous information on snow parameters and run-off. The applied methods and sensors of the in situ component as well as hydrological services were designed, developed and demonstrated in the framework of the business applications demonstration project SnowSense (2015–2018), which was co-funded by the European Space Agency (ESA). The SnowSense service mainly targets snow hydrological applications and is based on three pillars, including (i) a newly developed SnowSense in situ snow monitoring stations based on GNSS signals, (ii) EO products of the snow cover extent and information if and where the snow is dry or wet, and (iii) an integrated physically-based hydrological model.

The in situ and EO information are used to assimilate the input and the parameters of the applied hydrological model PROMET (Processes of Mass and Energy Transfer) [22] to calculate SWE, snowmelt onset, and river run-off in catchments as spatial layers. Those data layers contain the relevant information for flood forecasts and hydropower plant management, particularly for so far non- or sparsely equipped catchments in remote areas. Within the project demonstration phase, we validated the GNSS in situ snow stations and the first run-off results of the combined approach were already provided as an operational service for a commercial hydropower plant company and the administration of the island of Newfoundland, Canada, being our first demo users.

2. Materials and Methods

2.1. The SnowSense Concept

The overall aim of the demonstration project SnowSense was to build up an integrated service for run-off and hydropower assessment and forecast related to snow cover dynamics in remote areas. The idea of the service concept encloses the integration of in total three components for the provision of snow hydrological information. The service is based on the integration of (i) GNSS-based in situ measurements, (ii) EO monitoring and (iii) a hydrological model, to generate SWE, snowmelt and run-off products for remote areas in a temporal and spatial resolution, related to user's operation defaults. We chose one day as temporal and 1 km as spatial resolution. Furthermore, it is possible to provide estimations and forecasts on hydropower generation (Figure 1).

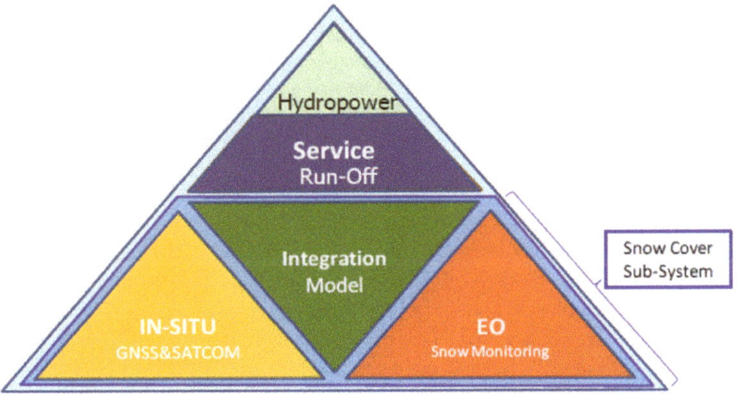

Figure 1. The modular concept of SnowSense, integrating satellite technologies like Global Navigation Satellite System (GNSS), satellite communication and Earth Observation (EO) as the basis for a service.

In the following subchapters, the applied Canadian test sites and the description, development and assimilation of the three basic components of the snow cover sub-system as well as the entire service are presented.

2.2. SnowSense Testsites in Canada

2.2.1. The Island of Newfoundland

The main region of interest for the development of the SnowSense service was the remote location of the island of Newfoundland (97,000 km^2), which is a province of Newfoundland and Labrador, Canada (Figure 2). The altitudinal gradient of this island ranges between 0 m and 800 m a.s.l. Newfoundland is primarily characterized by having a subarctic and a humid continental climate, with an annual precipitation ranging from 900 mm on the Northern Peninsula to 1700 mm in the southwestern region of the island [23,24]. This environment, together with a low population, provides the island of Newfoundland ideal conditions for taking advantage of using run-off for hydropower and building up reservoirs. However, the entire region has so far an insufficient instrumentation with only two functioning SWE measurement stations [25].

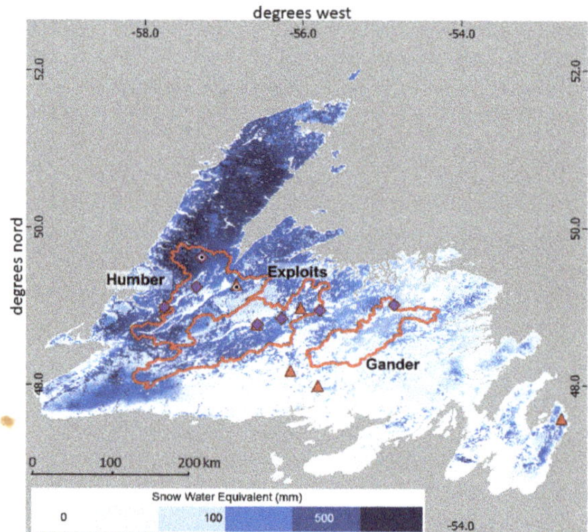

Figure 2. Overview on locations of SnowSense in situ stations (red triangles) and run-off stations (purple diamonds), within the Humber, Exploits, and Gander catchments in Newfoundland, Canada. Catchments are outlined with a red line. The locations of the two existing snow water equvalent (SWE) stations are indicated as black dots. The background shows the modelled SWE, which was assimilated by in situ stations and Earth Observation (EO) for the 15 March 2018.

As shown in Figure 2, we had access to several run-off gauges to validate our combined and assimilated hydrological service. In this study, we focus on the Humber and the Exploits catchments (Figure 2), which are of of most interest for the operations of the users. As marked in Figure 2, we set-up seven SnowSense in situ stations in Newfoundland, where two are located in the Exploits catchment and two in the Humber catchment. The hydrology of the catchment of the Humber River is influenced by the large lakes Deer Lake, Grand Lake, Sandy Lake and Hindis Lake. The water flow from Hindis to Grand Lake and to Deer Lake is regulated including the Deer Lake Power Plant (130 MW). Water flow from the Upper Humber area to Reidville is more or less natural.

The hydrology of the Exploits River catchment is strongly influenced by the Red Indian Lake (537 km^2), dividing the catchment into an upper and a lower part. The release of water from the lake at the Millertown Dam is very dynamic (55–700 m^3/s) and can be assessed from a downstream station.

In the framework of the demonstration project, we provided information gained from the SnowSense in situ stations and the combined EO and modelling approach as service to two dedicated demo users. Both demo users, namely Nalcor/NL Hydro and the Water Resources Management Division in the Department of Municipal Affairs and Environment (WRMD), are active in hydropower and flood forecast. During the project demonstration phase, they accessed, monitored and investigated the SnowSense products based on their specific requirements, whereof we present a selection of the results in this paper. For the two catchments as test sites, we set up the combined approach of in situ measurements, EO, and modelling.

2.2.2. The Forêt Montmorency NEIGE Site near Quebec

The Forêt Montmorency is located 70 km north of the city of Quebec and is the largest teaching and research forest in the world with an area of 412 km^2. It is open to the public and the University Laval operates in this forest the NEIGE test site (673 m a.s.l.). The Forêt Montmorency is drained by the Montmorency River and one of its tributaries, the Black River. The annual rainfall exceeds 1500 mm and in winter average snowfall is above 6 m.

As we had no possibility to validate SWE and LWC in the Humber and Exploit catchments in Newfoundland, we additionally had the opportunity to install a SnowSense in situ station at the study site NEIGE at Forêt Montmorency. At this location, numerous snow and meteorological instruments are tested and in operation. This encompasses, for example, the CS725 sensor, which was developed by Hydro Québec (Canada) in collaboration with Campbell Scientific (Logan, UT, USA) and which was used for comparison with the GNSS in situ measurements in this study [6]. Moreover, manual snow pit measurements, taken on a weekly basis, were available as further validation information on SWE. The CS725/GMON sensor was located 25 m next to the GNSS sensors, and the manual measurement are performed within a distance of less than 150 m. The site is a flat, sand/gravel surface area without significant vegetation. The close-by meteorological station site is in a distance of less than 500 m with an elevation difference less than 25 m. Since 2014, the University Laval is managing the NEIGE site, which is following the World Meteorological Organisation's (WMO) Canadian Solid Precipitation Experiment's (C-SPICE) objectives for solid precipitation measurements and comparisons of automatic instruments. During the winter season 2017/2018, several sensors and manual measurements from nine partners altogether took place at this location, which was a profound basis for comparing the results of our GNSS in situ station with other sensors.

2.3. In Situ Station Design and Setup

For the in situ component, we made use of information gained by signal attenuation and signal delay of GNSS signals passing through the snowpack to derive the snow cover properties SWE and LWC [8]. In general, it is possible to use any kind of GNSS signals such as the signals of the Global Positioning System (GPS), Galileo, GLONASS or Beidou for this approach. For this study, we used the freely-available GPS L1-band signals at a frequency of 1.57542 GHz. Recently, the additional usage of Galileo signals was integrated and will be applied in further studies [26]. The in situ GNSS sensor setup is in general composed of two static antennas as presented in Henkel et al. [11] and Koch et al. [8], whereof one antenna (GPS$_1$) was mounted on the top of a mast system to be permanently above the snow cover, and the other one (GPS$_2$) was placed on the ground before the first snow fall. In winter, when GPS$_2$ was covered by a layer of snow, the received GPS signals were markedly influenced by signal attenuation and time delay within the snowpack. For further information on the setup, the algorithms, the detailed GNSS signal processing steps as well as their validation at a high-alpine study site in Switzerland, we refer to [8,10,11].

A large focus regarding the development and the design of the SnowSense in situ sensor system for the study presented in this paper was to provide a practical and lightweight solution for the retrieval of continuous snow cover properties like SWE and LWC. The setup is in particular suitable for locations, which were so far not equipped with measurement sensors or which were difficult to access and not regularly visited by field observations. Having this purpose in mind, we designed, produced and tested an easily transportable solution consisting of a mast system with rigging. Regarding the deployment of the sensor system, it is practically possible to carry all sensor components in a large backpack, and two persons can install it within approx. two hours. Furthermore, the sensor system consists of two multi-GNSS sensors (u-blox, Thalwil, Switzerland), which receive GPS and Galileo signals, a processing unit capable for on-board processing of SWE and LWC as well as a power and a communication management unit. The latter two units are described in detail in [26]. The on-board processing is more power consuming than simply storing and transmitting raw data measurements. However, we promote the on-board processing to reduce the amount of data to be transmitted and to provide real-time data directly from the station to the client. The autonomous power supply for the operation and processing is provided by a combination of a solar panel and a battery pack and is controlled by the integrated power unit. As in many remote regions like in Newfoundland, the mobile network is not available, and we used an Iridium satellite communication (SatCom) module to transmit a daily or even sub-daily overview of the processed data. However, for regions with mobile networks, the communication unit can also be replaced by this technique. Figure 3a,b shows examples of some station components like the low-cost GNSS antennas and receivers as well as the microcontroller used for processing. Figure 3 gives an example of an installed sensor system and the solar panels on a 3 m high stable aluminum mast consisting of three pluggable segments.

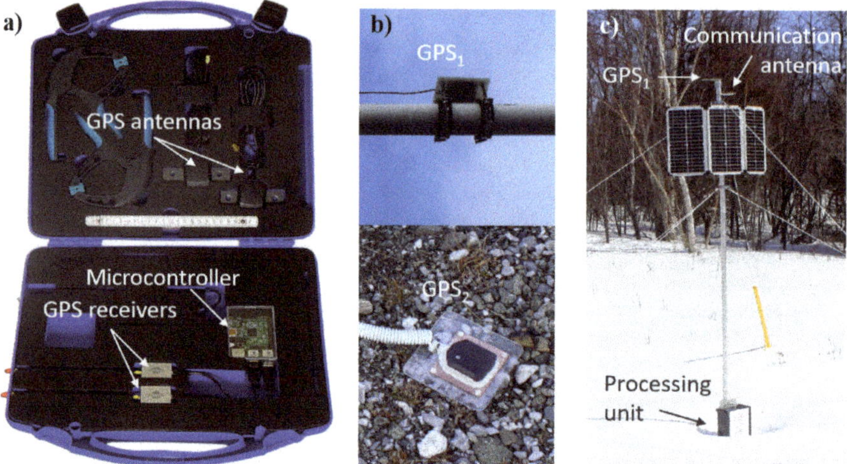

Figure 3. (**a**) demo suitcase with the applied low-cost LEA-M8T GPS receivers (u-blox, Thalwil, Switzerland), u-blox patch antennas and a small microcontroller and data storage; (**b**) examples of GPS antennas of GPS$_1$ mounted at a pole and of GPS$_2$ laid out on the ground before snowfall; (**c**) example of a SnowSense station including a mast system, a self-supplied power management (solar panels on top), a processing unit (including battery pack), and a communication module installed near Millertown, Newfoundland, Canada. GPS$_1$ and the communication antenna are mounted at the top of the mast and GPS$_2$ is covered by snow.

2.4. EO and Data Processing

For wet-snow detection from space, we used Sentinel-1A and -1B EO data in interferometric wide (IW) swath mode. IW mode is the main acquisition mode over land and satisfies the majority of service requirements. It acquires data with a 250 km swath at 5 m by 20 m spatial resolution [27]. The satellite data is automatically downloaded from Copernicus Data Hub (https://scihub.copernicus.eu/), and pre-processed using ESA Sentinel Toolbox for geometry and terrain corrections and it is spatially resampled on a 100 m by 100 m grid [28]. Afterwards, the data is radiometrically calibrated and corrected with regard to the elevation angle [29]. The wet-snow mapping is performed as proposed by Nagler and Rott [30] with a fixed threshold of -3 dB from a reference scene. Here, we are using a predefined reference data set with averaged standard backscatter values for different satellite geometries according to Appel et al. [29].

The results are binary wet snow maps with a resolution of 1 km by 1 km, containing three classes: wet snow, no information (which includes dry snow or snow free conditions), and no data. The maps were used in a further step for the assimilation process (see Section 2.6) and as additional visualization product, i.e., provided for the demo users in a password protected online portal (see Figure 4).

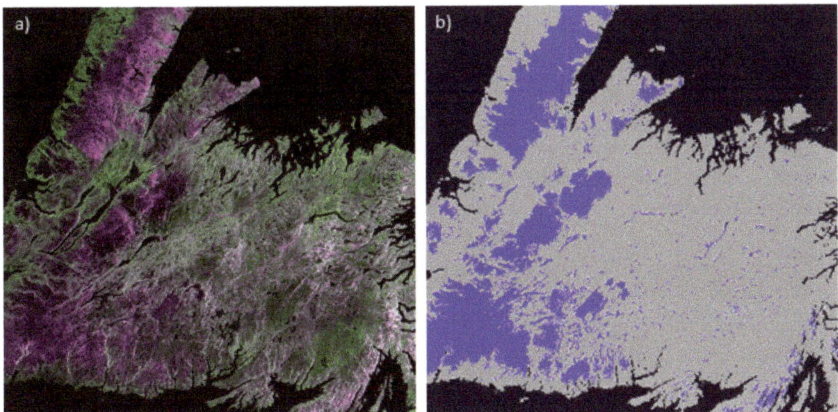

Figure 4. (**a**) Sentinel-1 IW mode composite for the Island of Newfoundland for 30 April 2017; (**b**) derived map of wet snow area (blue) from Sentinel-1 data.

2.5. Processes of Mass and Energy Transfer—PROMET Model Component

For the retrieval of the spatially distributed snow information and forecast capabilities, the hydrological part of the coupled land surface process model PROMET [22] was chosen. It is a raster-based model that has been developed to spatially simulate the elements of the land surface water balance including vegetation, soil moisture, snow cover, reservoirs, and river discharge. PROMET incorporates a four-layer soil model and considers lateral flows along hill slopes. Due to its raster-based architecture, the model allows for the assimilation of remote sensing data for distributed hydrological applications [31]. PROMET solves the water and energy balance for hourly time steps and calculates the run-off of river basins, while it strictly conserves mass and energy. It thus allows for the validation of the complete process chain, from rainfall over soil-moisture dynamics to vegetation controlled evapotranspiration and finally routed run-off, against measured discharges on the basin and sub-basin scale.

PROMET has successfully been applied for a variety of hydrological studies in medium-to large-sized watersheds [22,31–33]. Required input data consist of raster-based GIS information, characterizing the spatial distribution of land use patterns, soil types and topographic features. In addition, parameters describing the characteristics of the soil and land use or crop categories are supplied through tabular input.

For the application on the island of Newfoundland, we set up a basic model environment based on a freely available digital elevation model (DEM), from which the river network was automatically derived and lakes or reservoirs were inserted as structures. Land use information was derived from EO data sources (GlobCover), as soil map the Harmonized Soil Map of the World [34] is used. All spatial input data as well as the model grid itself have a resolution of 0.00833 degrees (<1 km^2). The parameterization of the catchment characteristics followed the internal standardized procedure developed by VISTA, purely based on GIS and EO data. As the PROMET model is physically based, it does not require a calibration. Further fine-tuning of parameters would be possible, but were, however, not conducted due to missing information e.g., on reservoir management and limited validation data.

PROMET can be driven by high resolution (1 h) time series of meteorological station network records, using the measured meteorological parameters. However, due to the specific characterization of our study area, only a sparse meteorological station network was available, which was not sufficient for our service target. Moreover, even in case of a sufficient number of meteorological stations, the amount of continuously provided information, e.g., on radiation and humidity or dew point are often missing or inaccurate. To overcome the lack of meteorological station data, we used global and regional climate model data as input, similar as presented by Zabel [35]. For the application in Newfoundland, we used data provided by Government of Canada as part of their High Resolution Deterministic Prediction System (HRDPS). This system is a set of nested limited-area model (LAM) forecast grids from the non-hydrostatic version of the Global Environmental Multiscale (GEM) model with a 2.5 km horizontal grid spacing.

The fields in the HRDPS high resolution data set are made available as GRIB2 for four times a day for the Pan-Canadian domain for a 48-h forecast period [36]. Those data sets are consistent in temporal (1 h) and spatial (2.5 km) resolution. We adapted the import routines for the meteorology and selected the appropriate parameters for the model forcing to perform calculations of the snow and run-off situation with hourly temporal resolution. Due to the architecture of PROMET, we are able to calculate any periods, starting from previously stored status and recovery files, and provide all information on the current snow and hydrological situation as spatial data sets in a 1 km by 1 km resolution with an hourly timestep and as tabled point data.

2.6. Assimilation

In general, and as, e.g., presented in [37–42], data assimilation is likely to improve the accuracy of the modelled snow parameters in a way that the modelled snow parameter or run-off results agree better with the real situation. For the SnowSense data assimilation, we used information gathered by the SnowSense GNSS in situ stations and the already existing online SWE measurement sites, as well as information from EO data. Regarding the in situ locations, we used the SWE and LWC (were available), regarding the EO sources, we used binary information whether the snow is dry or wet from Sentinel-1. All information was then compared twice a week with the non-assimilated output of SWE, LWC and the snow cover extent simulated by the PROMET model. In case of significant differences between simulated and observed values, assimilation runs for the last calculation period had been triggered. To adjust the model output to the observed in situ and EO information, the following options were available [43]: Adjustments of (i) the precipitation gain for the amount of snow/solid precipitation, (ii) the critical temperature for the transition of liquid and solid precipitation, and (iii) the albedo parameter for the short wave energy fluxes controlling the ablation dynamic. The sequence of the assimilation steps is following a decision tree based work-flow. After adjusting the model parameterization, the model is re-calculating the required output parameters for an improved agreement between model and reality. The adjustment of i,ii and iii is performed spatially. During the winter 2017/2018 phase, the assimilation process was performed partly manual. For a future service application, we will establish a more automated process.

3. Results and Discussion

In the first part of this section, we present the results and the validation of the GNSS in situ station for the well-equipped study site NEIGE at Forêt Montmorency. In the second part, we focus on the results and the validation of the hydrological service of the combined and assimilated run-off at, in total four, gauges of the Humber and Exploit catchments in Newfoundland.

3.1. Station Performance at the Forêt Montmorency NEIGE Site near Quebec

The SnowSense station was installed in October 2017 at the NEIGE site at Forêt Montmorency and was operational during the entire snow-covered winter season 2017/18. The power supply and the communication unit resisted the cold and windy environment without damage or failure. The station delivered continuous daily SWE and LWC measurements without any significant discontinuities (Figure 5a).

In general, the SWE values derived by the SnowSense GNSS in situ station are in good agreement with the provided reference measurements by the two GMON CS725 sensors and the manual snow pit measurements (Figure 5b).

The GMON sensor as well as the GNSS sensor are both non-destructive measurement methods and are largely capable of deriving SWE. Both sensors were already validated against other sensors like snow pillows and manual measurements (e.g., [7,11,44]). As stated by Choquette et al. [6], at the observed study site NEIGE, an average error of 18% between manual measurements and the GMON sensor is reasonable, and, for SWE levels less than 400 mm w.e., the estimation is inside the 5–10% range.

At the time the GNSS measurements started in early winter 2017, the GNSS-derived SWE as well as the measurements of the two GMON sensors lay in a similar range at approximately 100 mm w.e. At the end of April 2018, the maximum amount of SWE with approximately 500 mm w.e. was reached, which was indicated by the GNSS solution as well as the manual measurements and one GMON sensor. Comparing the two GMON sensors (blue solid and blue dashed line, Figure 5b), however, an offset of the SWE measurement of up to 30% occurred between them. This offset was low in the beginning of the time series and increased during the winter continuously. The reasons for this might originate in different sensor locations e.g., with slightly different wind conditions, and are still under discussion, but are out of the scope of this paper. The manual measurements were performed 16 times on a weekly basis during the snow season. On each day, three to four snow pits were analyzed. The resulting averaged SWE measurements from the snow pits lie well in between the range of the two GMON sensors. The SWE results of the SnowSense GNSS station (black solid line) follows in the beginning of the season the lower GMON sensor (CS725_TL, blue dashed line). Since 20 February, after a heavy precipitation event, the GNSS derived SWE jumps to the level of the GMON sensor with the higher SWE values (CS725_K, blue solid line). The coefficient of determination (R^2) between the GNSS data and the data from the GMON sensors is 0.53 for CS725_TL and 0.93 for CS725_K, respectively.

Throughout the entire season, the GNSS measurements agree very well with the manual measurements from snow pits including their minimum and maximum values. Here, the coefficient of determination (R^2) is 0.64. For all R^2 and root mean square errors (RMSE) errors of the validation study, we refer to Table 1.

Table 1. Statistical comparison of SWE derived by GNSS, two GMON sensors (CS725_K and CS725_TL), and manual snow pits on a weekly basis at the NEIGE testsite at Forêt Montmorency.

	RMSE [mm w.e.]	R^2
GMON CS725_K, manual	71.94	0.71
GMON CS725_TL, manual	91.38	0.15
GMON CS725_K, GNSS	35.89	0.93
GMON CS725_TL, GNSS	68.22	0.53
manual, GNSS	65.98	0.64

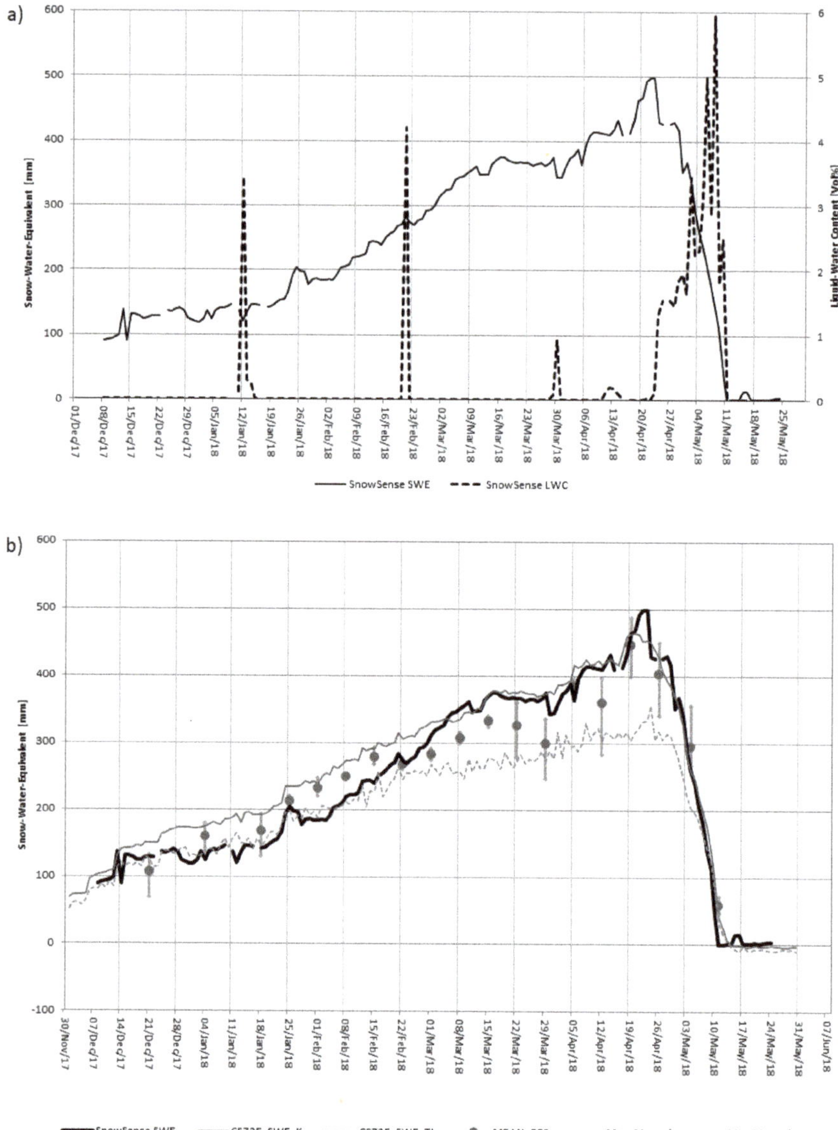

Figure 5. (**a**) daily GNSS-derived SWE and LWC values at the Forêt Montmorency NEIGE site 2017/2018; (**b**) daily GNSS-derived SWE values and reference measurements from two GMON CS725 sensors and manual snow pit measurements of mean, minima and maxima SWE at the Forêt Montmorency NEIGE site.

The range of the SWE validation results presented in this study for the Forêt Montmorency are in good accordance with previously conducted studies validating the GNSS in situ SWE component successfully at the high-alpine study site Weissfluhjoch in the Swiss Alps as presented in Henkel et al. [11] and Koch et al. [8]. One key result of their validation is a very good performance of the SWE determination capabilities of the in situ GNSS approach: the inter-comparison of the SnowSense GNSS station with a snow pillow and manual measurements show a very high agreement indicated by the values of $\mathbf{R}^2$ close to 1 [11]. The root mean square errors (RMSE) between the measurements from GNSS, snow pillow, and from snow pits fit very well; they lay in the range of 11–24 mm w.e. [11] but might be higher regarding wet snow conditions, e.g., due to an increase in GNSS signal attenuation (approx. 45 mm w.e.) [8].

Until mid-April 2018, the snowpack was predominantly dry at the study site Forêt Montmorency. In January, February and March, only single wet-snow events of up to two to three days occurred. The occurrence of wet snow presented by the GNSS-derived LWC is shown in Figure 5a. As no further sensors for a comparison with the GNSS-derived LWC were available at this site, we compared the GNSS-derived LWC with meteorological parameters. In general, the measured LWC is in a good temporal correspondence with rainfall events and goes along with warm air temperatures (Figure 6), which was also demonstrated in Koch et al. [10]. However, as shown in other previous studies, we were able to successfully validate the GNSS LWC measurements with other sensors like an upward-looking ground-penetrating radar and capacity probes at the study site Weissfluhjoch [45].

a)

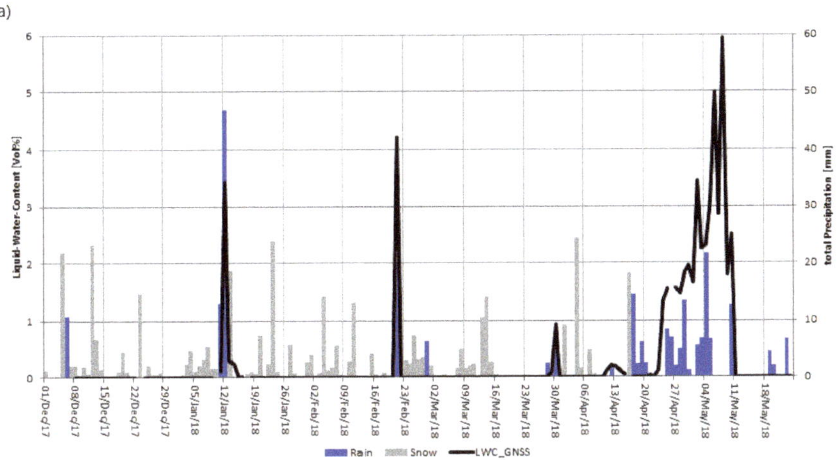

b)

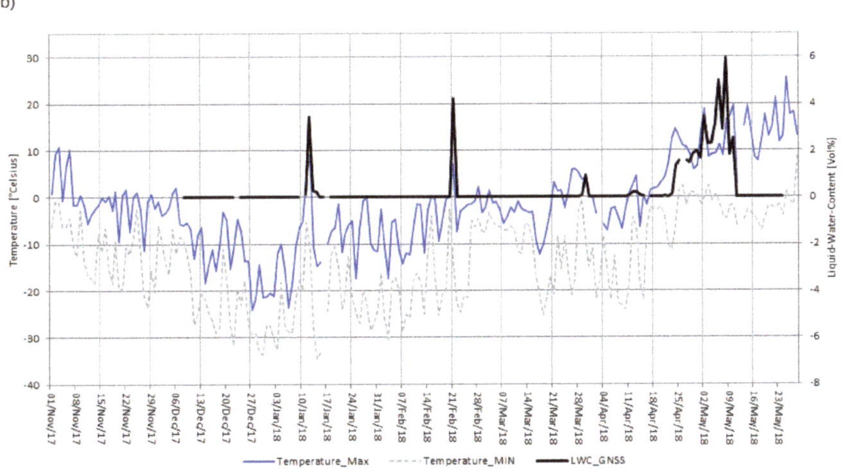

Figure 6. Daily GNSS-derived LWC values at the Forêt Montmorency NEIGE site for the winter season 2017/2018 related to (**a**) rain and snowfall events and (**b**) air temperature development. Temperature and precipitation measurements were conducted at the close-by provincial and federal station sites.

3.2. SnowSense Service for the Island of Newfoundland

For the entire SnowSense service based on the GNSS in situ measurements, EO and modelling, spatially-distributed maps of the SWE were generated and run-off estimates were derived for several river locations in the two main catchments Humber River and Exploit River. Figure 2 gives an example of a SWE map, which was modelled and assimilated for the entire island of Newfoundland for the 15 March 2018. Such a SWE map was provided for each day.

The simulated and assimilated run-off results were validated against four run-off gauge measurements in total. Measurements (raw data) are made available by the Water Resources Management Divisions. As for the catchments of the Humber River and the Exploits River, a strong interest in flood forecast was predominating; the run-off results are presented for the gauge stations at the lower parts of the two catchments.

3.2.1. Humber River

The run-off results for Humber Village for the winter season December 2017–May 2018 show very good agreement with the measured run-off reference values (Figure 7). Assimilation of the in situ information, mainly adjusting the solid precipitation, provide a perfect match of the volume of water. For the entire region, the amount of water stored as snow was perfectly represented by the model. The volume of total water released in the winter season 2017/18 was modelled by 95% (80% without assimilation). The coefficient of determination (R^2) reaches 0.90 at Humber Village. The results for the upper part of the catchment down to Reidville also show a very reasonable accordance between measured and modelled run-off behaviour (Figure 8). Due to invalid run-off measurements during the first flood peak in January 2018, no full analyses of the period were performed. However, the volume of total water released in up to Reidville was modeled by 90% (65% without assimilation). The coefficient of determination (R^2) reaches 0.80 at Reidville.

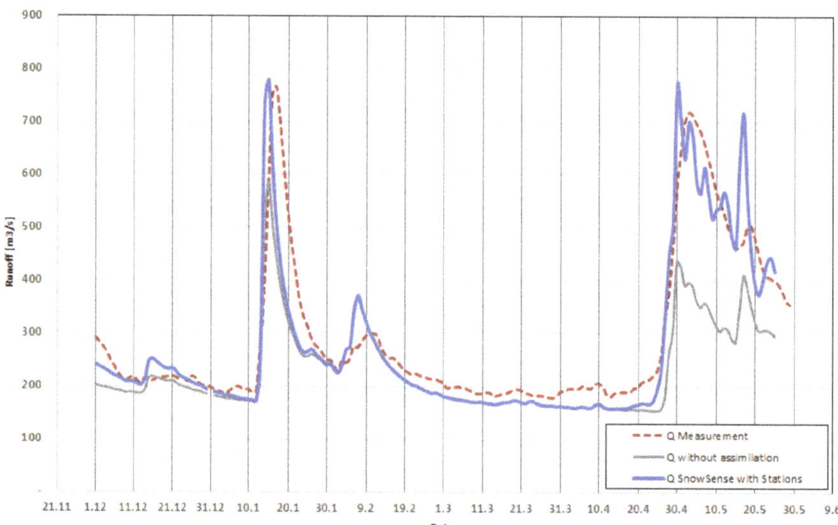

Figure 7. Daily run-off results from PROMET without (grey), with assimilation (blue) of stations and EO against the values (dashed red) measured at Humber Village (catchment: 7600 km^2) for the winter season 2017/18.

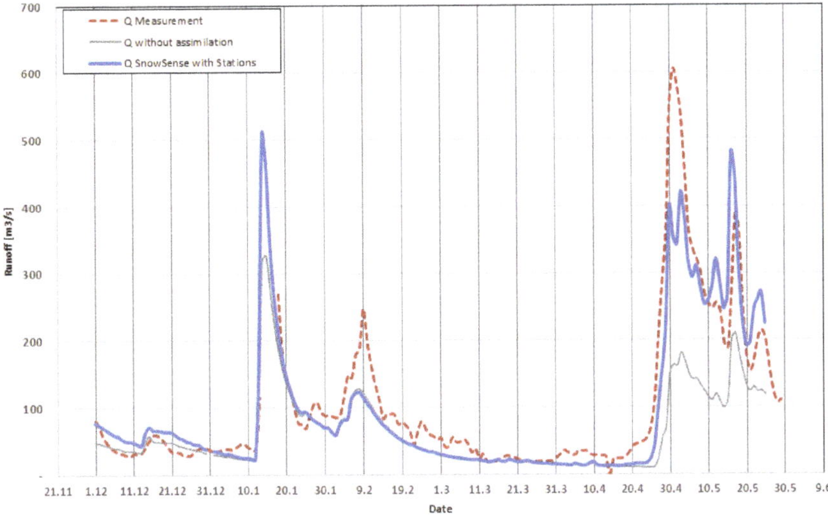

Figure 8. Daily run-off results from PROMET without (grey), with assimilation (blue) of stations and EO against the values (dashed red) measured at Reidville (catchment: 2100 km^2) for the winter season 2017/18.

3.2.2. Exploits River

The analyzed run-off results for the down-stream gauges "Below Noel Pauls Brook" and "Charlie Edwards Point" at the Exploits River show a very notable agreement regarding timing and volume with the measured values in the winter season from December 2017–May 2018 (Figures 9 and 10). Due to the strong influence of the controlled water release at Millertown Dam, and the not daily adapted model parameters during the demo, the results are a little below the ones for Humber River. The coefficient of determination (R^2) reaches 0.80 at Noel Paul Brook. For the downstream section of the Exploits River, after the conjunction with Badger River, the occurrence of river ice impeded the full evaluation of the results. Within the period of mid-December to late April, the measurement point was affected by ice jams, which resulted in an increased water level and therefore a false determination of the run-off. Due to the thawing in April, the flood peak from the snow melt could be well compared (Figure 10).

3.3. Advantages and Potential Limitations

Regarding the SnowSense in situ station, it is capable of measuring reliably SWE and LWC using freely available GNSS signals and low- cost GNSS sensors. Besides the derivation of SWE and LWC, Koch et al. [8] recently presented an approach to additionally derive snow height, which might even extend the range of application, not only for hydrological targets. A great advantage of the SnowSense in situ station is its light-weight design making an easy transportation and installation possible, which is highly valuable especially for remote and difficult to access areas. In total, only two people are needed for the set-up and all components can be carried in a big backpack. As the stations have an integrated on-board-processing module and satellite communication capabilities, the results can be transmitted (sub-)daily to the users. This makes the station autonomous and guarantees low maintenance. In general, the in situ station can either be used as a stand-alone component for snow cover property determination or can easily be integrated in the entire service encompassing the EO and modelling components.

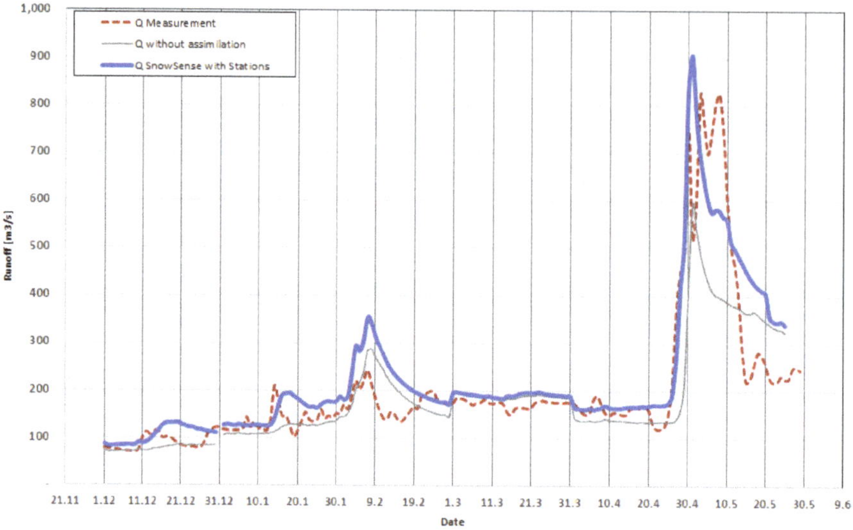

Figure 9. Daily run-off results from PROMET without (grey), with assimilation (blue) of stations and EO against the values (dashed red) measured at Noel Pauls Brook (catchment: 7400 km^2) for the winter season 2017/18.

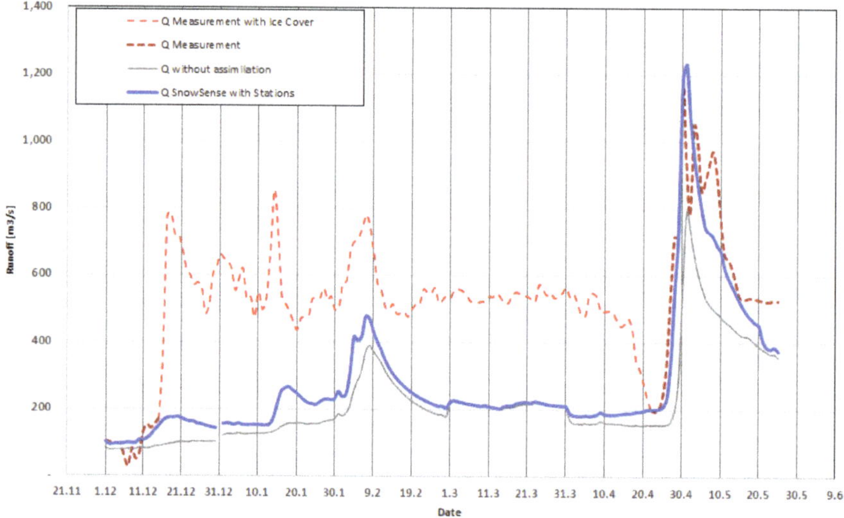

Figure 10. Daily run-off results from PROMET without (grey), with assimilation (blue) of stations and EO against the values (dashed red) measured at Charlie Edwards Point (catchment: 8850 km^2) for the winter season 2017/18. Due to river ice for the period mid-December to late April, the measurements have been separated into measurements with ice cover (normal dashed line) and without (bold dashed line).

As GNSS signals are globally available, the application of such in situ stations is potentially possible all over the world. However, a potential restriction of the in situ stations might be the availability of satellite reception in extreme locations, e.g., in narrow alpine valleys or in dense forests, with reduced GNSS signal reception. As presented in Lamm et al. [26], the integration of Galileo satellites besides GPS satellites increases satellite availability markedly, which increases also the availability in potentially difficult areas. Further studies will focus in more detail on, e.g., tilted terrain like avalanche-prone slopes, different climate and altitudinal ranges as well as more challenges on the station design regarding its protection from wildlife or its mounting on top of bare rocks and ice. Until now, we were able to test the GNSS SWE-derivation for quite huge amounts of snow with up to 1000 mm w.e. in high alpine regions [8]. In this study, we reached SWE values of up to 500 mm w.e. However, further studies on even more extreme amounts of snow and especially wet snow will be conducted in the future, as the limited operational time span of the demonstration project did not provide an opportunity to test the sensor performance for such extreme events performance. This is also true for further tests on extreme temperatures for the entire sensor hardware, which is designed for minimum temperatures of $-40\,°C$, though temperatures down to $-35\,°C$ were reached in December 2017 and January 2018 at the NEIGE testsite at Forêt Montmorency (Figure 6b).

Regarding the different SWE measurement techniques applied at the NEIGE study site at Forêt Montmorency, the GMON sensor is based on passive gamma rays, whereas the GNSS based measurements are based on electromagnetic waves. Both techniques are capable of deriving SWE in good accordance with standard measurement techniques like manual measurements or snow pillows [6,8,46]. Of course, slight differences in the derivation of SWE might occur considering these two relatively new sensors, however, as the sensors are not installed at the exact same place and are located up to 25 m from each other and have a distance (up to 150 m) to the manual measurements. The main differences might originate in different amounts of snow at each location, e.g., due to different wind effects and the different physical principles of the measurements.

Regarding the spatially distributed components EO and the hydrological model, it is another big advantage that both can also be potentially used worldwide and are often free of charge as, e.g., the Sentinel data. Of course, remote sensing products might be restricted in temporal and spatial resolution and face, depending on the wavelength and if the systems are active or passive, different limitations as, e.g., cloud cover or foreshortening effects. Therefore, it is often more difficult to apply EO in mountainous terrain. The applied hydrological model PROMET was already tested and validated for various applications for small and large catchments (e.g., [22,32,33,47]) and also globally (e.g., [48]) in different temporal and spatial resolutions. Although there are a few limitations in the model setup like, for example, the difficulty of implementing small-scale features regarding snow variability or the run-off generation in extreme alpine surroundings, the modelled output provides very good results for different scales, also in case of sparse input data as it is the case in Newfoundland. Especially in those remote areas with sparse data, real-time information and forecasts of run-off, fresh water availability, SWE, snow extent and the snowmelt onset can be significantly improved. Up to now, a limitation in our hydrological model setup is the lake ice formation. In general, SWE is calculated from the snow cover on the ground. However, until now, snow accumulation on top of frozen lakes is not implemented, although this SWE also contributes to the run-off after the onset of snow- and ice-melt. Moreover, we aim to describe river ice formation as an additional feature in the model since it can build up to ice dams with a subsequent banking up of water masses. These two model improvements will be implemented in the future, e.g. as suggested in [49,50].

The big advantage of the entire combined SnowSense service is that it picks up the advantages of all three components to deliver a reliable, assimilated product of snow and hydrology. In case data of one component is missing, the service can still rely on two other pillars and is therefore less vulnerable to data losses or other failures. The service can be used as entire combined system relying on the three pillars in situ measurements, EO, and hydrological modelling, but each of these pillars can also be applied as a modular stand-alone solution if desired. It therefore enhances and combines existing solutions and is, due to its modularity, a customer friendly approach.

3.4. Demo User Feedback

The feedback from the two demo users, Nalcor/NL Hydro and the Water Resources Management Division in the Department of Municipal Affairs and Environment (WRMD) of the Government of Newfoundland and Labrador, confirmed the significant importance of snow and run-off monitoring, which could be improved by using such a combined approach like the SnowSense service. They underlined that this is especially important for remote locations, as many regions in Newfoundland are only accessible by helicopter or snowmobiles and face limited access possibilities due to poor weather conditions. From their experience, standard in situ measurements sometimes fail, as the snowpack is often icy with a varying snow density and snow hardness, which makes it difficult to measure SWE with manual snow core equipment. Normally, manual snow surveys are performed three times per winter, which is expensive and is difficult to be completed under bad weather conditions, and causes labour safety risks. Existing automatic instrumentation (GMON sensors) is not providing enough information. The demo users state a need for accurate, reliable SWE information for each hydrological watershed and runoff forecasting that affects the real-time scheduling of hydrological assets and minimizes the use of thermal heat generation. From their point of view, a network of stations across the island of Newfoundland and Labrador would be the ideal scenario to give a province wide estimate of SWE for all users. In addition, they confirmed that SnowSense has the potential to fill a gap regarding the provision of spatial and temporal data at a high resolution by applying EO and modelling for spatial SWE maps of the entire region and run-off information at specific points of interest for real-time and forecasts. The received products matched with other sources of information, which they had for comparison and could even provide insights in hydrological processes. Both users stated that the in situ stations are competitive in their operation compared to other SWE monitoring technologies and therefore they have the potential to replace existing SWE monitoring stations like snow pillows or manual field observations.

In general, the feedback provided by the demo users was very positive, which encourages us in our further developments and improvements.

4. Conclusions

Within the ESA business applications demonstration project SnowSense (2015–2018), we successfully demonstrated a large scale snow hydrological monitoring service, by combining newly developed in situ stations based on signals of the Global Navigation Satellite System (GNSS), Earth Observation (EO) and hydrological modelling. With this combined approach, we present a reliable, and cost-efficient tool for the determination of snow cover properties like snow water equivalent (SWE), snow liquid water content (LWC), snow extent as well as run-off assessment, for real-time and forecast applications.

The GNSS in situ component was successfully applied and validated at the well-equipped study site NEIGE at Forêt Montmorency, Quebec, Canada. Furthermore, the entire SnowSense service providing modelled, in situ-, and EO-assimilated run-off was applied and validated at four run-off gauges within the Humber River and the Exploit River catchments on the island of Newfoundland, Canada.

The entire SnowSense service solution driven with an integrated numerical weather prediction (NWP) model for its application in this study in Newfoundland is capable of providing detailed knowledge on water stored as snow over large spatial scales. It is able to provide real-time and forecasted snow and run-off information and, if desired, also on reservoir status, which might be of great interest for hydropower plant operators. This information, which can be provided in various time steps, e.g., hourly up to daily, is especially needed in regions or catchments where in situ stations are only sparsely or non equipped catchments. The service is applicable at almost any location and was especially designed for remote locations, where access is limited and snow and run-off measurements were difficult up to now.

Within the project, the SnowSense service already reached a market dedicated design, based on the identification of potential customers (i.e., hydropower plants) and use cases (i.e., weather and climate observations, e.g., by national weather services).

Author Contributions: F.A., H.B. and W.M. conceived and developed the project idea; F.K., M.L., P.H. and F.A. were deeply involved in the development of the SnowSense algorithm, hardware development, and model integration; P.K. and H.B. were responsible for the PROMET model setup and the EO data assimilation. F.A. was the SnowSense project manager; F.A., P.H., F.K. and W.M. are involved in the patenting of the algorithm and the hardware; A.R., F.A. and F.K. wrote the paper.

Funding: The project was co-funded by the European Space Agency (ESA, 000113149/14/NL/AD) within the Business Application Demonstration Project SnowSense (2015–2018) (https://artesapps.esa.int/projects/snowsense-dp).

Acknowledgments: We thank the editors for their work and the two anonymous reviewers for their constructive comments. Florian Appel, Franziska Koch, Patrick Henkel, Markus Lamm, Philipp Klug and Anja Rösel were co-funded by the European Space Agency (ESA, 000113149/14/NL/AD) within the Business Application Demonstration Project SnowSense (2015–2018) (https://artesapps.esa.int/projects/snowsense-dp), which is gratefully acknowledged. The authors would like to thank all involved staff members of the demo users Nalcor/NL Hydro and the Department of Municipal Affairs and Environment, Water Resources Management Division (WRMD) of the Government of Newfoundland and Labrador, for their active support to the project, and for providing their run-off data. We thank the staff from University Laval for the opportunity to set up our sensor at the NEIGE site in Forêt Montmorency and for the opportunity to participate in the inter-comparison study. For the developed algorithm for the GNSS-based SWE and LWC determination as well as the invention of the in situ hardware, two patents are pending. SnowSense® is a registered trademark.

Conflicts of Interest: The authors declare no conflict of interest.

References

1. IPCC. *AR5: Climate Change 2013: The Physical Science Basis*; Cambridge University Press: Cambridge, UK, 2013.

2. Kinar, N.J.; Pomeroy, J.W. Measurement of the physical properties of the snowpack. *Rev. Geophys.* **2015**, *53*, 481–544. [CrossRef]

3. Pirazzini, R.; Leppänen, L.; Picard, G.; Lopez-Moreno, J.I.; Marty, C.; Macelloni, G.; Kontu, A.; von Lerber, A.; Tanis, C.M.; Schneebeli, M.; et al. European In-Situ Snow Measurements: Practices and Purposes. *Sensors* **2018**, *18*, 2016. [CrossRef] [PubMed]

4. Johnson, J.B.; Marks, D. The detection and correction of snow water equivalent pressure sensor errors. *Hydrol. Process.* **2004**, *18*, 3513–3525. [CrossRef]

5. Desilets, D.; Zreda, M.; Ferre, T.P.A. Nature's neutron probe: Land surface hydrology at an elusive scale with cosmic rays. *Water Resour. Res.* **2010**, *46*. [CrossRef]

6. Choquette, Y.; Lavigne, P.; Nadeau, M.; Ducharme, P.; Martin, J.P.; Houdayer, A.; Rogoza, J. GMON, a new sensor for snow water equivalent via gamma monitoring. In Proceedings of the International Snow Science Workshop, Whistler, BC, Canada, 21–27 September 2008.

7. Smith, C.D.; Kontu, A.; Laffin, R.; Pomeroy, J.W. An assessment of two automated snow water equivalent instruments during the WMO Solid Precipitation Intercomparison Experiment. *Cryosphere* **2017**, *11*, 101–116. [CrossRef]

8. Koch, F.; Henkel, P.; Appel, F.; Schmid, L.; Bach, H.; Lamm, M.; Prasch, M.; Schweizer, J.; Mauser, W. Retrieval of snow water equivalent, liquid water content and snow height of dry and wet snow by combining GPS signal attenuation and time delay. *Water Resour. Res.* **2018**, submitted.

9. Larson, K.; Gutmann, E.; Zavorotny, V.; Braun, J.; Williams, M.; Nievinski, F. Can we measure snow depth with GPS receivers? *Geophys. Res. Lett.* **2009**, *36*, L17502. [CrossRef]

10. Koch, F.; Prasch, M.; Schmid, L.; Schweizer, J.; Mauser, W. Measuring Snow Liquid Water Content with Low-Cost GPS Receivers. *Sensors* **2014**, *14*, 20975–20999. [CrossRef] [PubMed]

11. Henkel, P.; Koch, F.; Appel, F.; Bach, H.; Prasch, M.; Schmid, L.; Schweizer, J.; Mauser, W. Snow Water Equivalent of Dry Snow Derived From GNSS Carrier Phases. *IEEE Trans. Geosci. Remote Sens.* **2018**, *56*, 3561–3572. [CrossRef]

12. Steiner, L.; Meindl, M.; Fierz, C.; Geiger, A. An assessment of sub-snow GPS for quantification of snow water equivalent. *Cryosphere* **2018**, *12*, 3161–3175. [CrossRef]

13. Hall, D. *Remote Sensing of Ice and Snow*; Springer: New York, NY, USA, 2012.

14. Tedesco, M. *Remote Sensing of the Cryosphere*; Wiley: Hoboken, NJ, USA, 2014.

15. Malenovsky, Z.; Rott, H.; Cihlar, J.; Schaepman, M.E.; Garcia-Santos, G.; Fernandes, R.; Berger, M. Sentinels for science: Potential of Sentinel-1, -2, and -3 missions for scientific observations of ocean, cryosphere, and land. *Remote Sens. Environ.* **2012**, *120*, 91–101. [CrossRef]

16. Hall, D.K.; Riggs, G.A.; Salomonson, V.V.; DiGirolamo, N.E.; Bayr, K.J. MODIS snow-cover products. *Remote Sens. Environ.* **2002**, *83*, 181–194. [CrossRef]

17. Lopez, P.; Sirguey, P.; Arnaud, Y.; Pouyaud, B.; Chevallier, P. Snow cover monitoring in the Northern Patagonia Icefield using MODIS satellite images (2000–2006). *Glob. Planet. Chang.* **2008**, *61*, 103–116. [CrossRef]

18. Sirguey, P.; Mathieu, R.; Arnaud, Y. Subpixel monitoring of the seasonal snow cover with MODIS at 250 m spatial resolution in the Southern Alps of New Zealand: Methodology and accuracy assessment. *Remote Sens. Environ.* **2009**, *113*, 160–181. [CrossRef]

19. Kim, E.; Gatabe, C.; Hall, D.; Newlin, J.; Misakonis, A.; Elder, K.; Marshall, H.P.; Hiemstra, C.; Brucker, L.; De Marco, E.; et al. Overview of SnowEx Year 1 Activities. In Proceedings of the SnowEx Workshop, Longmont, CO, USA, 8–10 August 2017. Abaliable online: https://ntrs.nasa.gov/search.jsp?R=20170007518 (acceesed on 9 January 2019);

20. Cline, D.W.; Bales, R.C.; Dozier, J. Estimating the spatial distribution of snow in mountain basins using remote sensing and energy balance modeling. *Water Resour. Res.* **1998**, *34*, 1275–1285. [CrossRef]

21. Immerzeel, W.; Droogers, P.; de Jong, S.; Bierkens, M. Large-scale monitoring of snow cover and runoff simulation in Himalayan river basins using remote sensing. *Remote Sens. Environ.* **2009**, *113*, 40–49. [CrossRef]

22. Mauser, W.; Bach, H. PROMET–Large scale distributed hydrological modelling to study the impact of climate change on the water flows of mountain watersheds. *J. Hydrol.* **2009**, *376*, 362–377. [CrossRef]

23. Banfield, C.E.; Jacobs, J.D. Regional Patterns of Temperature and Precipitation for Newfoundland and Labrador during the past Century. *Can. Geogr.* **1998**, *42*, 354–364. [CrossRef]

24. Environment Canada. The Climate of Newfoundland. Available online: http://atlantic-web1.ns.ec.gc.ca/climatecentre/default.asp?lang=En&n=83846147-1 (accessed on 9 January 2019).

25. Bobba, A.G.; Lam, D.C.L.; Kay, D.; Ullah, W. Interfacing a hydrological model with the RAISON expert system. *Water Resour. Manag.* **1992**, *6*, 25–34. [CrossRef]

26. Lamm, M.; Koch, F.; Appel, F.; Henkel, P. Estimation of Snow Parameters with GPS and Galileo. In Proceedings of the International Symposium ELMAR, Zadar, Croatia, 16–19 September 2018.

27. European Space Agency (ESA). SENTINEL-1 SAR User Guide. Available online: https://sentinel.esa.int/web/sentinel/user-guides/sentinel-1-sar (accessed on 9 January 2019).

28. Appel, F.; Bach, H.; Heege, T.; de la Mar, J.; Siegert, F.; Rücker, G. APPS4GMES–Development of operational products and services for GMES a bavarian initiative. In Proceedings of the ESA Living Planet Symposium, Edinburgh, UK, 9–13 September 2013.

29. Appel, F.; Bach, H.; Loew, A.; Ludwig, R.; Mauser, W. Monitoring and Modeling of the Snow Cover Dynamic in Southern Germany-Capabilities of Optical and Microwave Remote Sensing. In Proceedings of the 25th Annual Symposium of the European Association of Remote Sensing Laboratories (EARSeL), Porto, Portugal, 6–11 June 2005.

30. Nagler, T.; Rott, H. Retrieval of wet snow by means of multitemporal SAR data. *IEEE Trans. Geosci. Remote Sens.* **2000**, *38*, 754–765. [CrossRef]

31. Mauser, W.; Schädlich, S. Modelling the spatial distribution of evapotranspiration on different scales using remote sensing data. *J. Hydrol.* **1998**, *212–213*, 250–267. [CrossRef]

32. Ludwig, R.; Mauser, W. Modelling catchment hydrology within a GIS based SVAT-model framework. *Hydrol. Earth Syst. Sci.* **2000**, *4*, 239–249. [CrossRef]

33. Strasser, U.; Mauser, W. Modelling the spatial and temporal variations of the water balance for the Weser catchment 1965–1994. *J. Hydrol.* **2001**, *254*, 199–214. [CrossRef]

34. Nachtergaele, F.; van Velthuizen, H.; Verelst, L. Harmonized World Soil Database. In Proceedings of the 19th World Congress of Soil Science, Soil Solutions for a Changing World, Brisbane, Australia, 1–6 August 2010.

35. Zabel, F.; Mauser, W. 2-way coupling the hydrological land surface model PROMET with the regional climate model MM5. *Hydrol. Earth Syst. Sci.* **2013**, *17*, 1705–1714. [CrossRef]

36. Government of Canada. HRDPS Data in GRIB2 Format. Available online: https://weather.gc.ca/grib/grib2_HRDPS_HR_e.html (accessed on 9 January 2019).

37. Andreadis, K.M.; Lettenmaier, D.P. Assimilating remotely sensed snow observations into a macroscale hydrology model. *Adv. Water Resour.* **2006**, *29*, 872–886. [CrossRef]

38. Nagler, T.; Rott, H.; Malcher, P.; Müller, F. Assimilation of meteorological and remote sensing data for snowmelt runoff forecasting. *Remote Sens. Environ.* **2008**, *112*, 1408–1420. [CrossRef]

39. Magnusson, J.; Gustafsson, D.; Hüsler, F.; Jonas, T. Assimilation of point SWE data into a distributed snow cover model comparing two contrasting methods. *Water Resour. Res.* **2014**, *50*, 7816–7835. [CrossRef]

40. Sensoy, A.; Schwanenberg, D.; Sorman, A.; Akkol, B.; Montero, R.; Uysal, G. Assimilating H-SAF and MODIS Snow Cover Data into the Conceptual Models HBV and SRM. In Proceedings of the Conference on EGU, Vienna, Austria, 27 April–2 May 2014.

41. Jörg-Hess, S.; Griessinger, N.; Zappa, M. Probabilistic Forecasts of Snow Water Equivalent and Runoff in Mountainous Areas. *J. Hydrometeorol.* **2015**, *16*, 2169–2186. [CrossRef]

42. Helmert, J.; Sensoy Sorman, A.; Alvarado Montero, R.; De Michele, C.; de Rosnay, P.; Dumont, M.; Finger, D.C.; Lange, M.; Picard, G.; Potopova, V.; et al. Review of Snow Data Assimilation Methods for Hydrological, Land Surface, Meteorological and Climate Models: Results from a COST HarmoSnow Survey. *Geosciences* **2018**, *8*, 489. [CrossRef]

43. Appel, F.; Bach, H.; Ohl, N.; Mauser, W. Provision Of Snow Water Equivalent From Satellite Data and the Hydrological Model PROMET Using Data Assimilation Techniques. In Proceedings of the IGARSS 2007 International Geoscience and Remote Sensing Symposium, Barcelona, Spain, 23–28 July 2007.

44. Wright, M.; Kavanaugh, J.; Labine, C. Performance Analysis of GMON3 Snow Water Equivalency Sensor. In Proceedings of the 79th Annual Western Snow Conference, Stateline, NV, USA, 18–21 April 2011.

45. Schmid, L.; Koch, F.; Heilig, A.; Prasch, M.; Eisen, O.; Mauser, W.; Schweizer, J. A novel sensor combination (upGPR-GPS) to continuously and nondestructively derive snow cover properties. *Geophys. Res. Lett.* **2015**, *42*, 3397–3405. [CrossRef]

46. Choquette, Y.; Ducharme, P.; Rogoza, J. CS725, an accurate sensor for the snow water equivalent and soil moisture measurements. In Proceedings of the International Snow Science Workshop, Grenoble, France, 7–11 October 2013; pp. 931–936.

47. Mauser, W.; Klepper, G.; Zabel, F.; Delzeit, R.; Hank, T.; Putzenlechner, B.; Calzadilla, A. Global biomass production potentials exceed expected future demand without the need for cropland expansion. *Nat. Commun.* **2015**, *6*, 8946. [CrossRef] [PubMed]

48. Mauser, W.; Prasch, M. *Regional Assessment of Global Change Impacts: The Project GLOWA-Danube*; Springer: New York, NY, USA, 2015.

49. Lindenschmidt, K.E.; Sydor, M.; Carson, R.W. Modelling ice cover formation of a lake–river system with exceptionally high flows (Lake St. Martin and Dauphin River, Manitoba). *Cold Reg. Sci. Technol.* **2012**, *82*, 36–48. [CrossRef]

50. Shen, H.T. Mathematical modeling of river ice processes. *Cold Reg. Sci. Technol.* **2010**, *62*, 3–13. [CrossRef]

 *geoscience* 

Article

Analysis of QualitySpec Trek Reflectance from Vertical Profiles of Taiga Snowpack

Leena Leppänen * and Anna Kontu

Finnish Meteorological Institute, Space and Earth Observation Centre, Tähteläntie 62, 99600 Sodankylä, Finland; anna.kontu@fmi.fi
* Correspondence: leena.leppanen@fmi.fi; Tel.: +358-40-670-7133

Received: 20 September 2018; Accepted: 1 November 2018; Published: 6 November 2018

Abstract: Snow microstructure is an important factor for microwave and optical remote sensing of snow. One parameter used to describe it is the specific surface area (SSA), which is defined as the surface-area-to-mass ratio of snow grains. Reflectance at near infrared (NIR) and short-wave infrared (SWIR) wavelengths is sensitive to grain size and therefore also to SSA through the theoretical relationship between SSA and optical equivalent grain size. To observe SSA, the IceCube measures the hemispherical reflectance of a 1310 nm laser diode from the snow sample surface. The recently developed hand-held QualitySpec Trek (QST) instrument measures the almost bidirectional spectral reflectance in the range of 350–2500 nm with direct contact to the object. The geometry is similar to the Contact Probe, which was previously used successfully for snow measurements. The collected data set includes five snow pit measurements made using both IceCube and QST in a taiga snowpack in spring 2017 in Sodankylä, Finland. In this study, the correlation between SSA and a ratio of 1260 nm reflectance to differentiate between 1260 nm and 1160 nm reflectances is researched. The correlation coefficient varied between 0.85 and 0.98, which demonstrates an empirical linear relationship between SSA and reflectance observations of QST.

Keywords: near-infrared reflectance; specific surface area; spectrometer; snow microstructure

1. Introduction

The microstructure of snow is an important parameter for the modelling of microwave emission and optical reflectance [1–3], and it is therefore also important for remote sensing applications. However, a parameter describing all snow properties, including size, shape, bonding, and the orientation of snow grains, is not simple to define. The parameters most often used for that purpose are traditional grain size [4], correlation length [3,5], optical grain size [6,7], and specific surface area (SSA) [8,9]. Several methods exist to measure these directly or indirectly, but this study concentrates only on the reflectance-derived, in-situ methods.

Snow reflectance at near infrared (NIR) and short-wave infrared (SWIR) wavelengths, especially over 1000 nm, is dominated by grain size [6]. Optical effective grain size is defined as the diameter of spheres having equal optical properties compared with the original grains [6,10]. SSA is defined as the surface area of the air–ice interface per unit mass (unit $m^2\ kg^{-1}$) [9,11]. Optical grain size and SSA are related by Equation (1):

$$SSA = \frac{6}{\rho_{ice}\, D_0} \tag{1}$$

where ρ_{ice} is the density of ice and D_0 is the optical grain size, which is derived based on equations in [9,12] as presented by [13]. Grain shape has been shown to affect the reflectance-derived SSA in many wavelengths [10,14,15], and [16] presented a modelling study concerning the influence of grain shape on albedo-derived SSA, resulting in 20–25% error.

DUal Frequency Integrating Sphere for Snow SSA (DUFISSS) was presented by [13]. It uses 1310 nm and 1550 nm lasers and an integrating sphere to measure hemispherical reflectance from the snow sample surface, which was converted to SSA [13,17]. The error of the SSA measurement was 10% based on [13]. The measurement method is similar to that of the IceCube, which is described in Section 2.2. The Profiler of Snow Specific Surface area Using SWIR reflectance Measurement (POSSSUM) and the Alpine Snow Specific Surface Area Profiler (ASSSAP) [18] are also based on the same principle. A contact spectroscopy method is presented by [7] to observe the optical grain diameter from a vertical profile of snowpack based on reflectance measurements of a band near 1030 nm by using a FieldSpec FR and attached modified Contact Probe (ASD Inc., Longmont, CO, USA). The derivation of optical grain size was based on an ice absorption model by [2]. The optical grain size and traditional grain size had poor correlation with less robust results for rounded grains, which is assumed to relate to the effect of grain shape. A review of past field experiments, where reflectance or albedo is compared to some of those parameters describing snow microstructure, is presented next.

Near infrared photography at 850–1000 nm is a method to observe SSA from reflectance as presented by [19]. The correlation between reflectance and SSA was reported to be 0.9 and the inaccuracy of SSA 15%. A clear correlation between hemispherical reflectance derived from measured nadir SWIR spectral reflectance of snow from FiedSpec FR 350–2500 nm (ASD Inc., Longmont, CO, USA) and SSA measured with Brunauer Emmett Teller (BET) analysis [9,20] was presented by [8]. Additionally, they did not find any effect from grain shape on the results. A correlation between grain size and a ratio of 1160 nm and 1260 nm hemispherical reflectance is showed by [21]. The accuracy of average grain size estimation was presented to be ±0.2 mm. Spectral albedo was measured with the Autosolexs instrument (400–1050 nm) to estimate SSA [22]. The accuracy of SSA depended on the solar zenith angle and the leveling of the instrument, and error was presented to be 15%. The SSA derived from Autosolexs had poor correlation with ASSSAP measurements; however, the wavelength and vertical resolutions of the instruments were different. The Automatic Spectro-Goniometer is presented for hemispherical–directional reflectance measurements [23]. The measurements were compared with Discrete Ordinates Radiative Transfer Program for a Multi-Layered Plane-Parallel Medium (DISORT)-modelled reflectance by using spheres with radii equal to the surface-area-to-volume ratio derived from stereological analysis, resulting in model underestimation with a maximum mean root-mean-square error (RMSE) of 0.09% for 1300 nm. A study where albedo was modelled with DISORT based on SSA measured with DUFISSS resulted in 1.1% difference to measured albedo with FieldSpec pro JR (ASD Inc., Longmont, CO, USA) [24]. It is estimated that 1% error in albedo corresponds approximately with 15 m^2 kg^{-1} error in SSA [16]. In addition, the optical diameter of snow grains with data derived from near infrared photography or FieldSpec (ASD Inc., Longmont, CO, USA) spectral albedo or reflectance observations is researched by [25–29]. Grain size or size distribution was measured traditionally or from image-processing of microphotography.

The presented methods from previous studies are primarily intended for measuring reflectance from the surface of the snowpack, and they are not suitable for measuring vertical profiles except for contact spectroscopy with the Contact Probe and near-infrared photography. Since solar radiation is subject to variability, originating from, for example, zenith angle and cloud cover, it is beneficial that the instrument used includes an internal light source for stable illumination. A newly developed instrument, QST (ASD Inc., Longmont, CO, USA), was tested for performing rapid measurements of vertical snow profiles. The hand-held instrument measures almost bidirectional reflectance with similar optical geometry to that of the Contact Probe. The instrument was tested on taiga snowpack in the Arctic Space Centre of the Finnish Meteorological Institute in Sodankylä in northern Finland. QST has been used previously for other purposes such as the detection of ion concentrations in soil [30] and heavy metal pollution in soil [31]. Snow measurements are more challenging due to the deeper penetration depth of the radiation, the fragile structure of the snowpack, and melting of the snow.

Previous results of SSA measurements from a taiga snowpack in Sodankylä include the comparison of SSA and optical grain size to modelling results. The optical grain size derived from SSA measurements of IceCube was compared to SNOWPACK [32–34] modelled optical grain size in [35,36]. SNTHERM [37,38], SNICAR [39], and MOSES [40] modelled optical grain sizes were compared to IceCube-derived optical grain size by [37]. Moreover, the spectral reflectance of a taiga snowpack has previously been measured with a portable Field Spec pro JR spectrometer in the Sodankylä area and a similar mast-based spectrometer located at the same area as the measurements performed for this study [41,42]. In addition, measurements of bidirectional reflectance factor have been performed with a goniospectropolariphotometer [43,44].

The aim of this study was to test the suitability of the QST instrument for measuring the reflectance of snow, and the main conclusion of the study was that an empirical relationship between SSA and QST reflectance exists. The paper has the following structure: Section 2 presents field measurements and methods, results are presented in Section 3, and discussion is made in Section 4.

2. Materials and Methods

Snow measurements have been performed at the Intensive Observation Area (IOA) in Sodankylä in northern Finland since 2006 (Figure 1) for the calibration, validation, and development of remote sensing instruments and interpretation algorithms. Vegetation in the cleared area among sparse pine forests consists of lichen, moss, heather, crowberry, and lingonberry, whose growth rate is approximately 0.4 cm per year in the snow pit area [45]. Typically, the average maximum snow depth (~80 cm) occurs in late March. The average air temperature is below 0 °C from November to April, and the average wind speed is low (1–2 m s^{-1}) in the area as described in [35]. Several automated instruments measuring snow, soil, radiation, and meteorological parameters are installed in the IOA, in addition to manually recorded snow pit measurements [35]. The data set for this study was taken from snow pit measurements made in spring 2017 (22 February, 7 March, 16 March, 21 March, and 3 April) with clear or partially cloudy sky conditions.

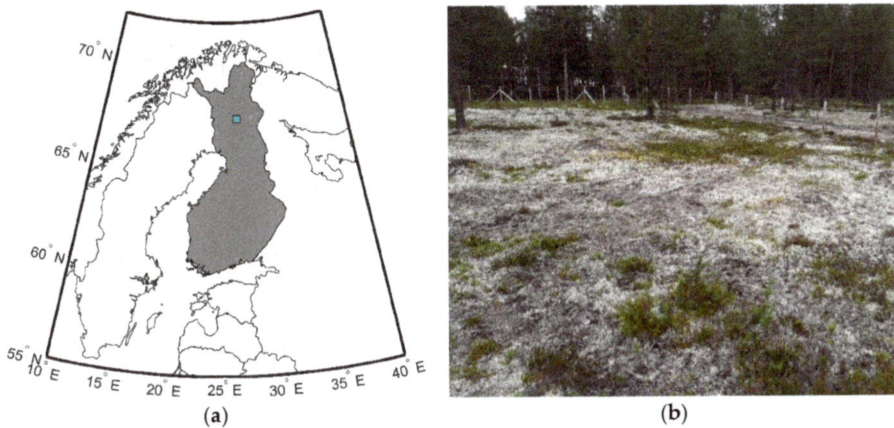

Figure 1. (a) Location of Sodankylä in northern Finland; (b) snow pit measurement site in Sodankylä.

2.1. Layers, Traditional Grain Size, and Density

Horizontal layers of the snowpack were defined manually with a brush and toothpicks. Layers were defined according to visual appearance, grain type, grain size, hardness, and wetness. Macrophotography-based, traditional grain size, the largest extent of an average grain [4], was estimated from macro-photographs taken against a 1 mm reference grid. The density profile was measured every 5 cm with a 10 cm × 10 cm × 5 cm rectangular sampler and digital scale with ±1 g accuracy. More detailed information of the methods is presented by [35].

2.2. SSA with IceCube

The reflectance and SSA was observed with IceCube manufactured by A2 Photonic Sensors, Grenoble, France. IceCube measures the hemispherical reflectance of a 1310 nm laser from the snow sample surface (Figure 2a). A photodiode observes diffuse radiation from the integrating sphere originating from the laser and reflected from the sample. The snow sample is collected with a spatula (or a specific tool), packed into the sample holder to reach a minimum density of 200 kg m^{-3}, and the surface is levelled (Figure 2b). The sample holder is 2.5 cm high and 6 cm in diameter. Software provided by the manufacturer converts the observed voltage value to reflectance using the calibration results. The instrument is calibrated for every measurement occasion using reference plates of different reflectivity. The calibration curve is fitted to the six calibration measurements. Determination of SSA from the reflectance is based on DISORT modelling and depends also on optical parameters of the integrating sphere [13]. In this study, the SSA samples were taken in 3 cm intervals from the vertical profile of the snowpack. A more detailed description of the IceCube measurement procedure is presented in [35] and a discussion on measurement errors is presented in [46].

Additional testing of the measurement accuracy occurred on 21 March 2017. The measurement of each sample from the profile was repeated three times with IceCube by rotating the sample in the azimuth direction between measurements. In addition, three samples were taken from both the surface layer and the depth hoar layer to test the effect of the sampling procedure and packing density. Those samples were also weighed to determine the sample density.

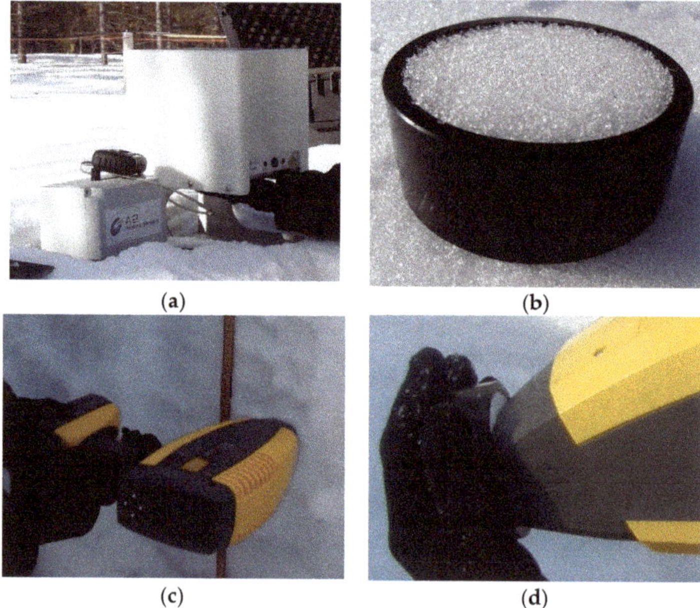

Figure 2. (**a**) The IceCube measurement; (**b**) the IceCube sample; (**c**) the QualitySpec Trek (QST) measurement from the snowpack vertical profile; (**d**) the QST measurement from sampled snow.

2.3. Reflectance with QualitySpec Trek

QST is a portable spectrometer manufactured by ASD Inc., Longmont, CO, USA (Figure 2c,d). QST measures the almost bidirectional reflectance of NIR and SWIR radiation at 350–2500 nm wavelengths with a spectral resolution of 9.8 nm at 1400 nm. The instrument has three detectors: 350–1000 nm (512-element silicon array), 1001–1785 nm (InGaAs photodiode), and 1786–2500 nm (InGaAs photodiode). The instrument also has an internal light source and internal gray scale reference for optimization and wavelength calibration. The light source is a Quartz Tungsten Halogen bulb with a color temperature of 2870 K $\pm$33 K. Illumination and viewing geometry is presented in Figure 3. The window of QST is approximately 1 cm in diameter, and the whole window is illuminated by the internal light source. The window and materials inside the instrument are designed to minimize specular reflections. The angle from the instrument window to the light source is 55° and the angle to the fiber optic is 78°. The fiber optic cable has a field of view of 25°, which means that the field of view is 0.82 cm on the window of the instrument. The Contact Probe [7] has the same measurement geometry as QST. Additional calibration is made before every measurement occasion with a separate white reference plate. The reference plate is attached in front of the instrument window with magnetic plugs. It is important that the white reference plate is clean and that the manufacturer provides information for cleaning and replacing the plate. Full-spectrum dark reference is measured also during the start-up with an internal shutter, the light source is turned off, and the white reference is plugged. The dark reference (background) value is subtracted from raw data prior to the reflectance calculation. The sample count averaging time can be set to 1, 2, 5, or 10 s. The instrument scans the entire wavelength range 10 times per second and produces an average of those. The sample is set to physical contact with the instrument window (Figure 2c,d and Figure 3), so cleaning of the window between the measurements is therefore required. A measurement is stored by pushing a trigger button. Sound signals for starting and finishing the measurement notes when the instrument needs to be in stationary position with the (snow) sample. The most recent data is shown on the screen. Audio note recording is possible after the measurement. The instrument also stores coordinates and elevation from the internal GPS for each sample. Automatic and manual data storage options are available. The resulting reflectance is absolute reflectance (reflectance normalized with reference reflectance). Since the instrument is commercial, calibration data and data from the single spectrums before averaging are not available.

The measurement procedure included a white reference measurement, instrument setup (choosing how many times the wavelength range is scanned for averaging), and snow measurement (repeated). Cleaning of the window was rarely needed. Automatic data storage was used. The integration time for a measurement was set to 1 or 2 s. A longer integration time meant more averaged spectrums and therefore less noise for an acquisition. On the other hand, the instrument remained steadier in snow with a shorter integration time. Two procedures related to how the instrument can be used are (1) vertical reflectance profile from the snowpack of the snow pit wall and (2) reflectance from the snow surface without digging a snow pit. The measurements were made with the first procedure from the snowpack profile by pushing the instrument window steadily against the snow pit wall during the measurement (Figure 2c). The window had physical contact to the snow so that the distance between the snow and fiber optic cable was always constant and the measurement configuration did not change. The measurements were made at approximately 2–5 cm intervals from the vertical profile. Some of the IceCube samples were also measured with QST by pushing the window against the snow in the middle of the sample surface (Figure 2d).

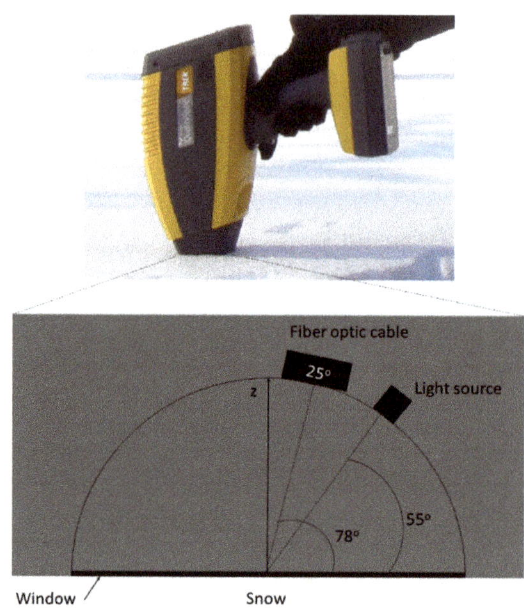

Figure 3. Illumination and viewing configuration. Fiber optic and light source have the same azimuth angle. Fiber optic cable has 25° field of view. Diameter of the window is approximately 1 cm.

2.4. Methods for Comparison of QST and IceCube Measurements

Reflectance profiles were measured with QST from both the snowpack and the sampled snow on 21 March and 3 April 2017. The two profiles were measured with different height intervals, and therefore both profiles were linearly interpolated for every centimeter (no extrapolation). A comparison presented in Section 3.3 was made only for the heights where both interpolated reflectance values existed.

The empirical relationship between SSA and reflectance was studied with 1160 and 1260 nm reflectance-dependent coefficient Q, which is the ratio of reflectance at the bottom of the ice absorption feature and the reflectance change in the ice absorption feature. The absorption coefficient of ice is larger at 1260 nm than at 1160 nm. Q is calculated with Equation (2) as

$$Q = \frac{R_{1260}}{R_{1160} - R_{1260}} \tag{2}$$

where R_{1160} is reflectance at 1160 nm and R_{1260} is reflectance at 1260 nm.

Similarly, interpolation was also needed to compare SSA and reflectance-derived Q from the same heights of snow. Linear interpolation without extrapolation was made for SSA when the reflectance profiles from the snowpack were studied. For 21 March, the values from the first of the three SSA measurements of the same sample were used. The comparison presented in Section 3.4 was made with interpolated SSA values from the same heights as the original reflectance profile measurements from the snowpack. Interpolation was not needed in the case of the reflectance profile from sampled snow because the same samples were measured with both instruments.

Clearly erroneous reflectance values measured with QST were removed from the analysis (>0.1 difference to the closest value without fitting to other values around). One of those was from the snowpack profile in 16 March (63 cm height), another one was from the snowpack profile in 21 March (70 cm height), and the two last ones were from the sampled snow in 21 March (48 and 51 cm height).

3. Results

The data set includes vertical profile measurements of reflectance, SSA, macrophotography-based traditional grain size, and density from various snow types with varying grain size and density in a taiga snowpack. The snow height of 0 cm was set to the ground surface. The height of the snow and air temperature for spring 2017 are presented in Figure 4. An overview of all data used in the study is presented in Figure 5.

3.1. Snowpack Properties

The height of snow increased before 21 March and melting had already started in 3 April, as shown in Figure 4. The air temperature was close to zero or positive (in °C) on 16 March, 21 March, and 3 April. Layer properties had temporal variations; however, there were some lasting similarities. The hard crust layer at approximately 30 cm height (red in Figure 5c) was observed in all snow pits. The traditional grain size was larger below than above that layer, which is typical, because grain growth is largest at the bottom due to the higher temperature gradient. Between 30–40 cm height of snow, the grain shape changed from round (pink in Figure 5c) to faceted (light blue in Figure 5c), and grain size increased simultaneously. In mid-winter, the bottom layers consisted of smaller grains than in the layers above those. Typically, traditional grain size was larger in faceted crystal and depth hoar layers than layers with rounded grains, as expected based on the temperature gradient. Density was lowest in the top 5–15 cm of the snowpack. Maximum density was found in the 5–15 cm bottom layer of the snowpack, and in the crust layer around 30–40 cm height. Based on those observations, the reflectance and SSA measurements were made at differing air temperatures and snow conditions with variating grain size, grain type, and density.

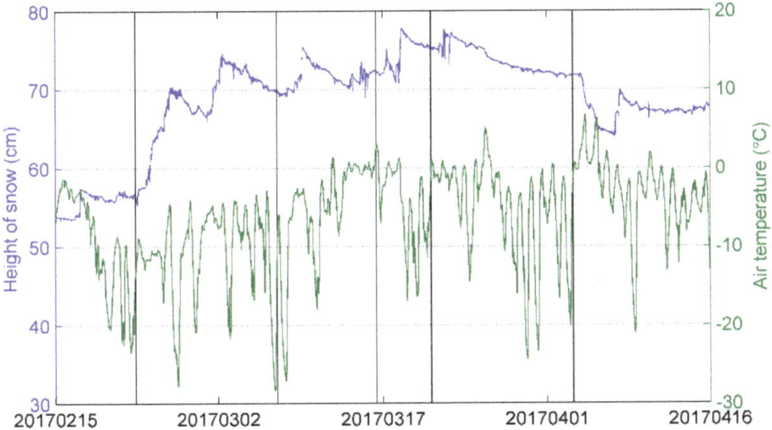

Figure 4. The height of snow (blue) and air temperature (green) measured at the Intensive Observation Area (IOA). Vertical black lines indicate the measurements made with QST.

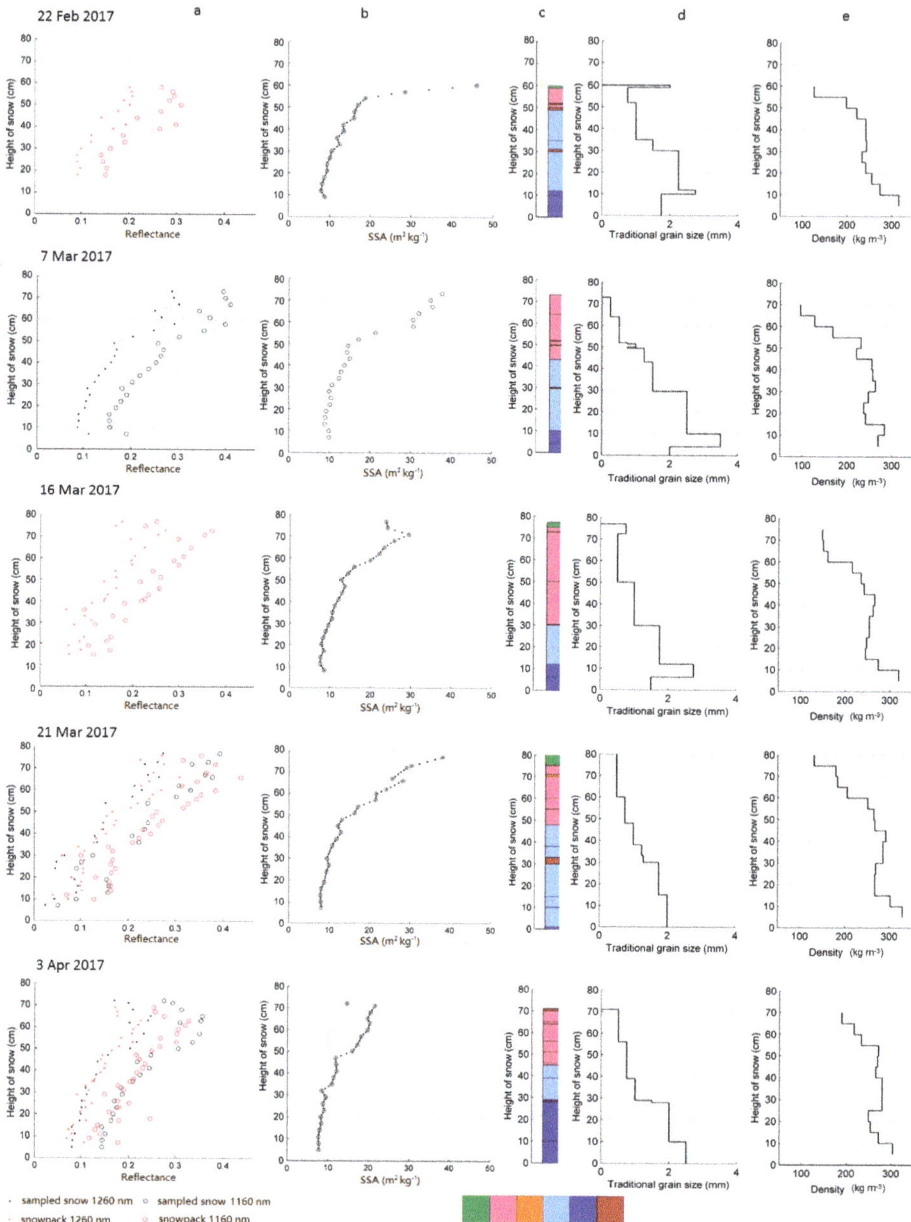

Figure 5. Reflectance from QST, specific surface area (SSA) from IceCube, grain type, macrophotography-based traditional grain size, and density profiles of snow are plotted in columns (**a**–**e**) for all dates. Original SSA data are marked with circles and interpolated values with dots in column (**b**). SSA for 21 March 2017 is plotted from the first measurement of three for every height. Grain type abbreviations are described in [4].

3.2. Repeatability of SSA Measurements

Repeatability of the SSA measurements with IceCube was tested to confirm the reliability of observations for the following analysis. First, the measurement method was tested by repeating measurements with the same sample; then, sampling was tested by sampling snow to different densities. The repeatability of the IceCube measurement method was tested on 21 March 2017. Every sample from the profile was measured three times with IceCube. The mean standard deviation of the SSA was 0.34 m^2 kg^{-1}, which means a relative standard deviation (RSD) of 2.0%. RSD is calculated by Equation (3)

$$RSD = 100\,\sigma/\mu \qquad (3)$$

where σ is the standard deviation and μ is the average. The repeatability of IceCube sampling was also tested on 21 March 2017. Precipitation particles in the surface layer and faceted crystals in the bottom of the snowpack were both sampled three times. The samples were taken next to each other so that snow was as homogenous as possible for all the samples. The samples were packed to different densities, measured with IceCube and weighed for density calculation. The average standard deviation was 0.65 and RSD was 2.85% (Table 1). Since the total RSD was low (<5%), SSA is relied on as a truthful value in the subsequent analysis.

Table 1. The SSA from three samples of precipitation particles at the surface and faceted crystals in the bottom of the snowpack. Standard deviations (STD) and relative standard deviations (RSD) from Equation (3) are calculated for both surface and bottom.

	Density (kg m^{-3})	SSA (m^2 kg^{-1})	STD	RSD (%)
	225.0	39.8		
Surface	309.9	41.6	1.04	2.53
	352.4	41.6		
	338.2	7.9		
Bottom	409.0	8.2	0.25	3.17
	394.9	7.7		
Average			0.65	2.85

3.3. Comparison of QST Reflectance Profiles

Reflectance profiles were measured with QST from both the snowpack and the sampled snow in 21 March and 3 April 2017. Both profiles were made next to each other at the same snow pit. The reflectance profiles were compared with three wavelengths (1160, 1260, and 1310 nm), which are used later in the study.

The comparison results showed a strong correlation between the reflectance profiles directly from the snowpack and the sampled snow with a correlation coefficient of 0.92–0.94 (Table 2), as was expected. Bias varied from −0.020 to 0.016 and RMSE varied from 0.022 to 0.040 (Table 2). According to the results, IceCube sampling had only a small effect on the reflective properties of snow (microstructure) at 1160, 1260, and 1310 nm wavelengths.

Table 2. Bias, root-mean-square error (RMSE) and correlation coefficient (*R*) for QST reflectances from the snowpack profile and the sampled snow.

Date	Wavelength (nm)	Bias	RMSE	*R*
	1160	−0.0204	0.040	0.92
21 March 2017	1260	−0.0202	0.031	0.94
	1310	−0.0230	0.034	0.93
	1160	0.0160	0.033	0.92
3 April 2017	1260	0.0055	0.022	0.92
	1310	0.0068	0.023	0.92

3.4. Empirical Relationship between SSA and Reflectance

Previously shown relationships between snow microstructural parameters and albedo or hemispherical reflectance are described in the Section 1. Similarly, we hypothesized that reflectance measured with QST could have an empirical connection to the SSA. The empirical relationship between SSA and reflectance was studied with 1160 and 1260 nm reflectance-dependent coefficient Q (see Section 2.4). Examples of the measured reflectance are presented in Figure 6, where the upper part of an ice absorption feature was located close to 1160 nm and bottom of it close to 1260 nm. The correlation between SSA and Q was high with correlation coefficients between 0.85 and 0.98 (Table 3). Linear fits are presented for measurements from the snowpack and the sampled snow in Figure 7. The fit is better for the QST measurements from the snowpack (Figure 7a). Single outliers existed for observations in 21 March from sampled snow, which are visible in the scatter plot of Figure 7b. There is approximately 0.2 bias in the linear fits between QST measurements from the snowpack and from the sampled snow. The results prove that an empirical relationship between SSA and reflectance exists, although measurement of light, new snow and fragile depth hoar is challenging.

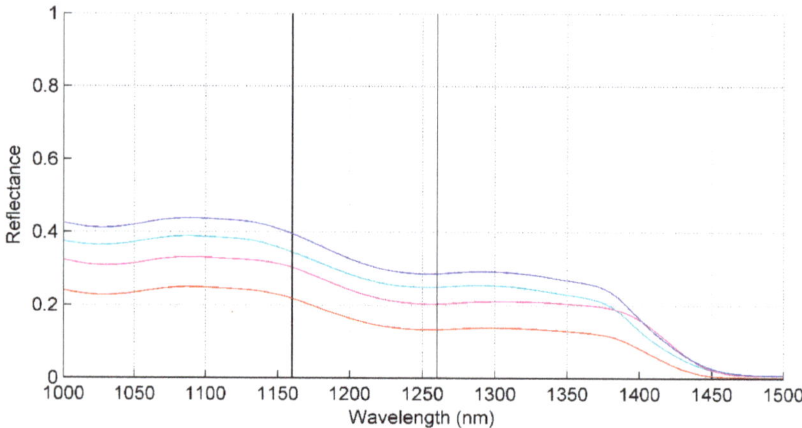

Figure 6. Example of reflectances measured on 7 March 2017 at heights of 73 cm (blue), 64 cm (cyan), 52 cm (magenta), and 34 cm (red). Vertical lines indicate wavelengths 1160 nm and 1260 nm.

Table 3. Correlation coefficient (R) between SSA and Q for the QST measurements from the snowpack profile and profile of the sampled snow.

Date	Profile	R
22 February 2017	Snowpack	0.95
7 March 2017	Sampled Snow	0.98
16 March 2017	Snowpack	0.89
21 March 2017	Sampled Snow	0.89
21 March 2017	Snowpack	0.94
3 April 2017	Sampled Snow	0.96
3 April 2017	Snowpack	0.85

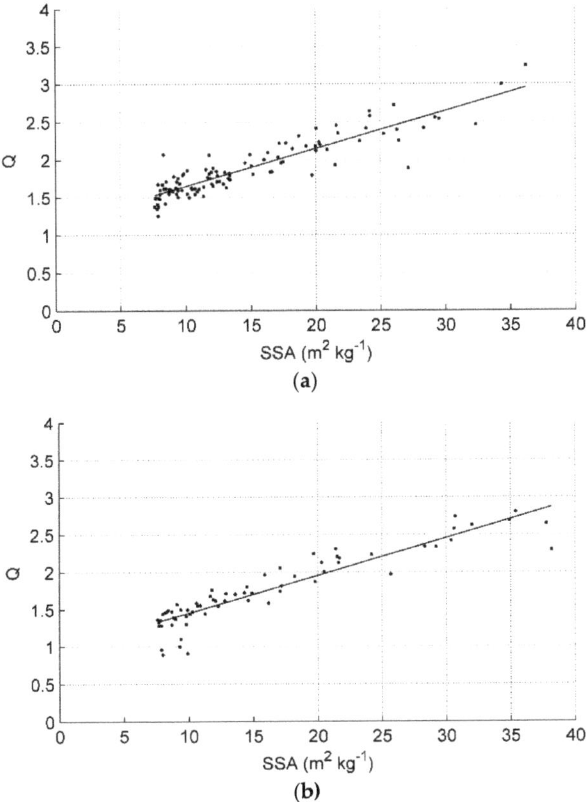

Figure 7. Scatter plot of Q with Equation (2) (*y*-axis) and SSA (*x*-axis) from the QST measurements made (**a**) directly from the snowpack profile and (**b**) the measurements from the sampled snow. Lines fitted to the points are marked.

4. Discussion

Snow microstructure is an important parameter for microwave and optical remote sensing of snow. Field observations are needed in the development and validation of retrieval algorithms. Therefore, simple and accurate novel measurement methods are required. In this study, the newly developed QST instrument for reflectance measurements was tested on taiga snow, and the empirical relationship between QST reflectance and the microstructural parameter SSA was defined.

We tested the measurement accuracy of SSA to confirm the repeatability of the measurement method. The accuracy was tested by repeating IceCube measurements (several measurements from the same sample) and repeating IceCube sampling (several samples from the same height of snow). The first one resulted in an error (relative standard deviation) of 2.0% and the second one resulted in an error of 2.9%. However, the repeated sampling was made for two of the most difficult types of snow to measure with IceCube with a relatively small set of observations. The total error was below 5%, and measurements are therefore considered repeatable in the taiga snow conditions. We hypothesized that snow sampling for IceCube measurements might change the microstructure of snow. However, the correlation of the two QST reflectance profiles (one directly from the snowpack and one from the sampled snow) was high, with correlation coefficients of 0.92–0.94. Based on this result, we assume that the IceCube sampling procedure does not remarkably affect the optical properties and microstructure of the snow in the sample. Analysis made from the IceCube and the QST measurements resulted in a strong correlation between SSA and Q, where Q is the ratio of reflectance at the bottom of the absorption feature (1260 nm) and the reflectance change in the absorption feature (between 1160 and 1260 nm). The reflectance profiles measured from both the snowpack and the sampled snow were used to calculate the correlation coefficients, which had values 0.85–0.98. In Section 1, the presented correlations of instruments measuring reflectance or albedo to SSA measurements are around 0.9. Compared to those values, the correlation between SSA and Q is approximately similar or better. Lower values of Q from the sampled snow compared to observations from the snowpack (bias of 0.2) originate likely from absorption by the black sample holder. The outlier values of the sampled snow on 21 March originate most likely from the melting of the snow samples due the longer measurement procedure, since the samples were measured first three times with IceCube, then weighed and finally measured with QST. Typically, the sampled snow was measured with QST directly after one IceCube measurement.

The QST measurements contain error arising from instrument metrics, the observer, and the environmental conditions. As a heavy instrument (2.5 kg), QST is difficult to hold stably in a stationary position with physical contact to the snow when it is not possible to lean the instrument on snow. This is the case with newly fallen surface snow with a low density (around 100 kg m^{-3}) and depth hoar layer snow, which is coarse and therefore fragile. This could be avoided by using an appropriate, assembled tripod or another support structure. On the other hand, adjusting the height of the tripod between each measurement of a profile might slow down the measurements. Physical contact to the snow is required so this needs to be confirmed when using external support. However, dense and compact snow (rounded grains or faceted crystals with traditional grain size <1.25 mm and density >200 km m^{-3}) had no such problems. In surface snow, solar radiation penetrates the snowpack and possibly causes some degree of error in the measurements. This could be avoided by covering the snowpack, as presented in [7,19]. The sampled snow measurements with QST were made mostly in the shade of the snow pit. The effect of different external illumination conditions was not studied. Warming of the instrument window and plastic casing causes some melting of the snow, and probably some inaccuracy to the measurements, especially during clear skies and positive air temperature conditions. Definition of the exact point of measurement in the snow profile is difficult, with the accuracy being approximately ±2 cm. The instrument size limits its ability to perform measurements close to the ground, so that the bottom of the snowpack below approximately 10 cm is difficult to measure. The clearly erroneous reflectance measurements, which were removed from the analysis, were expected to originate from the wetness of the snow or scattering of the radiation to outside of the field of view.

QST has similar optical geometry to the Contact Probe, and it uses the same range of wavelengths. However, QST is a compact, hand-held instrument while the Contact Probe is attached to a spectrometer. The hand-held instrument has no external spectrometer, laptop, and connecting cables, so QST is therefore fast and simple to use compared to the Contact Probe. In addition, audio notes are possible to record with QST. However, the Contact Probe has less direct contact with snow and the light source is further from the snow, which may reduce additional warming and melting of the snow. Reflectance from the Contact Probe is successfully used for calculation of optical grain size with the Nolin–Dozier model [7]. Therefore, we assume that derivation of optical grain size or SSA from the QST reflectance could be possible with a lookup table in the future, since the empirical relationship between SSA and reflectance has been found. However, it would require a more comprehensive set of observations and proper testing of the penetration depth of radiation and the lost portion of scattered radiation, where field experiments with both QST and the Contact Probe would be beneficial.

Author Contributions: Conceptualization, L.L. and A.K.; Formal analysis, L.L.; Investigation, L.L.; Methodology, L.L. and A.K.; Visualization, L.L.; Writing—original draft, L.L.; Writing—review & editing, L.L. and A.K.

Funding: Work of Leena Leppänen has been partly undertaken with funding from the Vilho, Yrjö and Kalle Väisälä Foundation of the Finnish Academy of Science and Letters.

Acknowledgments: We thank the personnel of Arctic Space Centre of Finnish Meteorological Institute for performing the manual snow measurements. We thank ASD Goetz instrument support program for the availability of the QualitySpec Trek instrument.

Conflicts of Interest: The authors declare no conflict of interest.

References

1. Pulliainen, J.; Grandell, J.; Hallikainen, M. HUT snow emission model and its applicability to snow water equivalent retrieval. *IEEE Trans. Geosci. Remote Sens.* **1999**, *37*, 1378–1390. [CrossRef]
2. Nolin, A.W.; Dozier, J. A hyperspectral method for remotely sensing the grain size of snow. *Remote Sens. Environ.* **2000**, *74*, 207–216. [CrossRef]
3. Mätzler, C. MATLAB functions for Mie scattering and absorption, version 2. *IAP Res. Rep.* **2002**, *8*, 1–24.
4. Fierz, C.; Armstrong, R.L.; Durand, Y.; Etchevers, P.; Greene, E.; McClung, D.M.; Nishimura, K.; Satyawali, P.K.; Sokratov, S.A. *The International Classification for Seasonal Snow on the Ground*; IHP-VII Technical Documents in Hydrology N°83, IACS Contribution N°1; UNESCO-IHP: Paris, France, 2009.
5. Debye, P.; Anderson, H.R.; Brumberger, H. Scattering by an inhomogeneous solid. II. The correlation function and its application. *J. Appl. Phys.* **1957**, *28*, 679–683. [CrossRef]
6. Warren, S.G. Optical properties of snow. *Rev. Geophys.* **1982**, *20*, 67–89. [CrossRef]
7. Painter, T.H.; Molotch, N.P.; Cassidy, M.; Flanner, M.; Steffen, K. Contact spectroscopy for determination of stratigraphy of snow optical grain size. *J. Glaciol.* **2007**, *53*, 121–127. [CrossRef]
8. Domine, F.; Salvatori, R.; Legagneux, L.; Salzano, R.; Fily, M.; Casacchia, R. Correlation between the specific surface area and the short wave infrared (SWIR) reflectance of snow. *Cold Reg. Sci. Technol.* **2006**, *46*, 60–68. [CrossRef]
9. Legagneux, L.; Cabanes, A.; Dominé, F. Measurement of the specific surface area of 176 snow samples using methane adsorption at 77 K. *J. Geophys. Res. Atmos.* **2002**, *107*. [CrossRef]
10. Grenfell, T.C.; Warren, S.G. Representation of a nonspherical ice particle by a collection of independent spheres for scattering and absorption of radiation. *J. Geophys. Res. Atmos.* **1999**, *104*, 31697–31709. [CrossRef]
11. Dominé, F.; Cabanes, A.; Taillandier, A.S.; Legagneux, L. Specific surface area of snow samples determined by CH4 adsorption at 77 K and estimated by optical microscopy and scanning electron microscopy. *Environ. Sci. Technol.* **2001**, *35*, 771–780. [CrossRef] [PubMed]
12. Kokhanovsky, A.A.; Zege, E.P. Scattering optics of snow. *Appl. Opt.* **2004**, *43*, 1589–1602. [CrossRef] [PubMed]
13. Gallet, J.C.; Domine, F.; Zender, C.S.; Picard, G. Measurement of the specific surface area of snow using infrared reflectance in an integrating sphere at 1310 and 1550 nm. *Cryosphere* **2009**, *3*, 167–182. [CrossRef]
14. Neshyba, S.P.; Grenfell, T.C.; Warren, S.G. Representation of a nonspherical ice particle by a collection of independent spheres for scattering and absorption of radiation: 2. Hexagonal columns and plates. *J. Geophys. Res. Atmos.* **2003**, *108*. [CrossRef]
15. Grenfell, T.C.; Neshyba, S.P.; Warren, S.G. Representation of a nonspherical ice particle by a collection of independent spheres for scattering and absorption of radiation: 3. Hollow columns and plates. *J. Geophys. Res. Atmos.* **2005**, *110*. [CrossRef]
16. Picard, G.; Brucker, L.; Fily, M.; Gallée, H.; Krinner, G. Modeling time series of microwave brightness temperature in Antarctica. *J. Glaciol.* **2009**, *55*, 537–551. [CrossRef]
17. Gallet, J.C.; Domine, F.; Dumont, M. Measuring the specific surface area of wet snow using 1310 nm reflectance. *Cryosphere* **2014**, *8*, 1139–1148. [CrossRef]
18. Arnaud, L.; Picard, G.; Champollion, N.; Domine, F.; Gallet, J.-C.; Lefebvre, E.; Fily, M.; Barnola, J.-M. Measurement of vertical profiles of snow specific surface area with a 1 cm resolution using infrared reflectance: Instrument description and validation. *J. Glaciol.* **2011**, *57*, 17–29. [CrossRef]
19. Matzl, M.; Schneebeli, M. Measuring specific surface area of snow by near-infrared photography. *J. Glaciol.* **2006**, *52*, 558–564. [CrossRef]
20. Brunauer, S.; Emmett, P.H.; Teller, E. Adsorption of gases in multimolecular layers. *J. Am. Chem. Soc.* **1938**, *60*, 309–319. [CrossRef]
21. Hyvärinen, T.; Lammasniemi, J. *Infrared Measurement of Grain Size and Melting Degree of Snow*; Infrared Technology XI; International Society for Optics and Photonics: San Diego, CA, USA, 1985; Volume 572, pp. 158–166.

22. Picard, G.; Libois, Q.; Arnaud, L.; Verin, G.; Dumont, M. Development and calibration of an automatic spectral albedometer to estimate near-surface snow SSA time series. *Cryosphere* **2016**, *10*, 1297–1316. [CrossRef]

23. Painter, T.H.; Dozier, J. Measurements of the hemispherical-directional reflectance of snow at fine spectral and angular resolution. *J. Geophys. Res. Atmos.* **2004**, *109*. [CrossRef]

24. Carmagnola, C.M.; Dominé, F.; Dumont, M.; Wright, P.; Strellis, B.; Bergin, M.; Dibb, J.; Picard, G.; Libois, Q.; Arnaud, L.; et al. Snow spectral albedo at Summit, Greenland: Measurements and numerical simulations based on physical and chemical properties of the snowpack. *Cryosphere* **2013**, *7*, 1139–1160. [CrossRef]

25. Langlois, A.; Royer, A.; Montpetit, B.; Picard, G.; Brucker, L.; Arnaud, L.; Harvey-Collardard, P.; Fily, M.; Goïta, K. On the relationship between snow grain morphology and in-situ near infrared calibrated reflectance photographs. *Cold Reg. Sci. Technol.* **2010**, *61*, 34–42. [CrossRef]

26. Aoki, T.; Aoki, T.; Fukabori, M.; Hachikubo, A.; Tachibana, Y.; Nishio, F. Effects of snow physical parameters on spectral albedo and bidirectional reflectance of snow surface. *J. Geophys. Res. Atmos.* **2000**, *105*, 10219–10236. [CrossRef]

27. Pirazzini, R.; Räisänen, P.; Vihma, T.; Johansson, M.; Tastula, E.M. Measurements and modelling of snow particle size and shortwave infrared albedo over a melting Antarctic ice sheet. *Cryosphere* **2015**, *9*, 2357–2381. [CrossRef]

28. Tanikawa, T.; Aoki, T.; Hori, M.; Hachikubo, A.; Abe, O.; Aniya, M. Monte Carlo simulations of spectral albedo for artificial snowpacks composed of spherical and nonspherical particles. *Appl. Opt.* **2006**, *45*, 5310–5319. [CrossRef] [PubMed]

29. Tanikawa, T.; Aoki, T.; Hori, M.; Hachikubo, A.; Aniya, M. Snow bidirectional reflectance model using non-spherical snow particles and its validation with field measurements. *EARSeL eProc.* **2006**, *5*, 137–145.

30. Peng, J.; Ji, W.; Ma, Z.; Li, S.; Chen, S.; Zhou, L.; Shi, Z. Predicting total dissolved salts and soluble ion concentrations in agricultural soils using portable visible near-infrared and mid-infrared spectrometers. *Biosyst. Eng.* **2016**, *152*, 94–103. [CrossRef]

31. Hu, B.; Chen, S.; Hu, J.; Xia, F.; Xu, J.; Li, Y.; Shi, Z. Application of portable XRF and VNIR sensors for rapid assessment of soil heavy metal pollution. *PLoS ONE* **2017**, *12*, e0172438. [CrossRef] [PubMed]

32. Bartelt, P.; Lehning, M. A physical SNOWPACK model for the Swiss avalanchewarning: Part I: Numerical model. *Cold Reg. Sci. Technol.* **2002**, *35*, 123–145. [CrossRef]

33. Lehning, M.; Bartelt, P.; Brown, B.; Fierz, C.; Satyawali, P. A physical SNOWPACK model for the Swiss avalanche warning. Part II: Snow microstructure. *Cold Reg. Sci. Technol.* **2002**, *35*, 147–167. [CrossRef]

34. Lehning, M.; Bartelt, P.; Brown, B.; Fierz, C. A physical SNOWPACK model for the Swiss avalanche warning: Part III: Meteorological forcing, thin layer formation and evaluation. *Cold Reg. Sci. Technol.* **2000**, *35*, 169–184. [CrossRef]

35. Leppänen, L.; Kontu, A.; Hannula, H.-R.; Sjöblom, H.; Pulliainen, J. Sodankylä manual snow survey program. *Geosci. Instrum. Methods Data Syst.* **2016**, *5*, 163–179. [CrossRef]

36. Kontu, A.; Lemmetyinen, J.; Vehviläinen, J.; Leppänen, L.; Pulliainen, J. Coupling SNOWPACK-modeled grain size parameters with the HUT snow emission model. *Remote Sens. Environ.* **2017**, *194*, 33–47. [CrossRef]

37. Sandells, M.; Essery, R.; Rutter, N.; Wake, L.; Leppänen, L.; Lemmetyinen, J. Microstructure representation of snow in coupled snowpack and microwave emission models. *Cryosphere* **2017**, *11*, 229. [CrossRef]

38. Jordan, R. *A One-Dimensional Temperature Model for a Snow Cover: Technical Documentation for SNTHERM.89*; Special Report 91-16; Cold Regions Research and Engineering Laboratory (U.S.): Hanover, NH, USA; Engineer Research and Development Center (U.S.): Vicksburg, MS, USA, 1991.

39. Flanner, M.G.; Zender, C.S. Linking snowpack microphysics and albedo evolution. *J. Geophys. Res.* **2006**, *111*, D12208. [CrossRef]

40. Essery, R.; Best, M.; Cox, P. *MOSES 2.2 Technical Documentation*; Technical Report, Hadley Centre Technical Note 30; Hadley Centre, Met Office: Exeter, UK, 2001; p. 31.

41. Salminen, M.; Pulliainen, J.; Metsämäki, S.; Kontu, A.; Suokanerva, H. The behaviour of snow and snow-free surface reflectance in boreal forests: Implications to the performance of snow covered area monitoring. *Remote Sens. Environ.* **2009**, *113*, 907–918. [CrossRef]

42. Meinander, O.; Kazadzis, S.; Arola, A.; Riihelä, A.; Räisänen, P.; Kivi, R.; Kontu, A.; Kouznetsov, R.; Sofiev, M.; Svensson, J.; et al. Spectral albedo of seasonal snow during intensive melt period at Sodankylä, beyond the Arctic Circle. *Atmos. Chem. Phys.* **2013**, *13*, 3793–3810. [CrossRef]

43. Peltoniemi, J.I.; Kaasalainen, S.; Näränen, J.; Matikainen, L.; Piironen, J. Measurement of directional and spectral signatures of light reflectance by snow. *IEEE Trans. Geosci. Remote Sens.* **2005**, *43*, 2294–2304. [CrossRef]

44. Peltoniemi, J.; Hakala, T.; Suomalainen, J.; Puttonen, E. Polarised bidirectional reflectance factor measurements from soil, stones, and snow. *J. Quant. Spectrosc. Radiat.* **2009**, *110*, 1940–1953. [CrossRef]

45. Leppänen, L.; Leinss, S.; Suokanerva, H. Growth of Forest Floor Vegetation Observed from Snow Depth Measurements in Sodankylä, Finland, Snow an Ecological Phenomenon, Smolenice, Slovakia, 19–21 September 2017. Available online: http://www.sbks.sk/smolenice/Leppanen%20et%20al.pdf (accessed on 5 November 2018).

46. Leppänen, L.; Kontu, A.; Vehviläinen, J.; Lemmetyinen, J.; Pulliainen, J. Comparison of traditional and optical grain-size field measurements with SNOWPACK simulations in a taiga snowpack. *J. Glaciol.* **2015**, *61*, 151–162. [CrossRef]

Article

Geometric Versus Anemometric Surface Roughness for a Shallow Accumulating Snowpack

Jessica E. Sanow [1,2], Steven R. Fassnacht [2,3,4,*], David J. Kamin [3,5], Graham A. Sexstone [6], William L. Bauerle [7] and Iuliana Oprea [8]

[1] Geosciences, Colorado State University, Fort Collins, CO 80523-1482, USA; Jessica.Sanow@colostate.edu
[2] Natural Resources Ecology Laboratory, Colorado State University, Fort Collins, CO 80523-1499, USA
[3] ESS-Watershed Science, Colorado State University, Fort Collins, CO 80523-1476, USA; kamind@gmail.com
[4] Cooperative Institute for Research in the Atmosphere, Colorado State University, Fort Collins, CO 80523-1375, USA
[5] Now with Xcel Energy, 1800 Larimer St, Denver, CO 80202, USA
[6] Colorado Water Science Center, U.S. Geological Survey, Lakewood, CO 80225, USA; sexstone@usgs.gov
[7] Horticulture and Landscape Architecture, Colorado State University, Fort Collins, CO 80523, USA; Bill.Bauerle@colostate.edu
[8] Mathematics, Colorado State University, Fort Collins, CO 80523, USA; Juliana@math.colostate.edu
*   Correspondence: Steven.Fassnacht@colostate.edu

Received: 2 November 2018; Accepted: 1 December 2018; Published: 6 December 2018

Abstract: When applied to a snow-covered surface, aerodynamic roughness length, z_0, is typically considered as a static parameter within energy balance equations. However, field observations show that z_0 changes spatially and temporally, and thus z_0 incorporated as a dynamic parameter may greatly improve models. To evaluate methods for characterizing snow surface roughness, we compared concurrent estimates of z_0 based on (1) terrestrial light detection and ranging derived surface geometry of the snowpack surface (geometric, z_{0G}) and (2) vertical wind profile measurements (anemometric, z_{0A}). The value of z_{0G} was computed from Lettau's equation and underestimated z_{0A} but compared well when scaled by a factor of 2.34. The Counihan method for computing z_{0G} was found to be unsuitable for estimating z_0 on a snow surface. During snowpack accumulation in early winter, z_0 varied as a function of the snow-covered area (SCA). Our results show that as the SCA increases, z_0 decreases, indicating there is a topographic influence on this relation.

Keywords: aerodynamic roughness length; terrestrial lidar; snow surface topography; wind profile; snow energy balance; snow accumulation

---

1. Introduction

In the Northern Hemisphere, a seasonal snowpack can cover over 50% of the land area with the snow surface often the interface between the atmosphere and the earth [1]. The roughness of a snow surface is an important control on air-snow heat transfer [2], and changes in the snow surface can have substantial effects on the energy balance at this interface. Snow is a complicated surface with rapidly evolving physical roughness characteristics due to changing atmospheric conditions, the metamorphism of snow crystals, melting and freezing processes and redistribution by wind, especially in open areas [3]. Roughness characteristics also influence the air-surface momentum transfer on the snowpack due to wind [4]. The changes in wind momentum can reduce the energy budget, influence the formation of roughness features, and affect the aeolian movement of snow [4]. Heat flux modeling has typically used the aerodynamic roughness length (z_0) as a static parameter, in hydrologic, snowpack, and climate models [5,6], with z_0 only varying as a function of land cover type. For example, the Community Land Model version 4.0 (CLM4; http://www.cesm.ucar.edu/models/ccsm4.0/clm/)

applies a single z_0 value of 2.4×10^{-3} m to all snow-covered surfaces. However, z_0 varies both spatially [7] and temporally [8], which may result in variable estimates of turbulent heat fluxes not captured by most models [9]. Wind velocity profile measurements are often used to calculate z_0 estimates [6], but there are a limited number of sites that measure the wind profile over a snowpack surface, making the spatio-temporal representation of z_0 challenging.

Millimeter-scale variations in snow-surface roughness features have been estimated from a black plate pushed partially into the snow [10–12], using two-dimensional photography, digital processing, and automated post-processing software [13–16]. Snow surface elevation data are now available over large areas at the resolution (±80 mm) of airborne light detection and ranging (lidar) [17–19], terrestrial laser scanner (TLS) (resolution of ±5 mm) [20–26], and photogrammetry [25]. Although most lidar and photogrammetry efforts have only focused on snow depth [26], only a few datasets have been used to evaluate snow surface roughness at the meter-scale or sub-meter scale [27,28]. However, few of these datasets have been applied to interpolate z_0 and create a digital elevation model of the snowpack surface for evaluating surface roughness [27]. Aerodynamic roughness length (z_0) has been estimated from the geometry of the snow surface [2,7,29–31]. However, this method is time consuming and typically only applicable over smaller scales [13]. Also, Fassnacht et al. [27] have identified potential errors with the different methods of computing z_0 from the geometry of the surface that result in values varying over 1–3 orders of magnitude and have suggested these methods need to be evaluated for varying scales, resolutions, and environments.

This study used TLS-derived surface geometry and vertical wind profile measurements to compare concurrent z_0 estimates for changing snow surface features of shallow snowpacks. Here, we asked the following questions: (1) How does the aerodynamic roughness length (z_0) vary spatially and temporally for a shallow snow environment? (2) How does z_0 estimated from geometric measurements (z_{0G}) compare to z_0 estimated from anemometric measurements (z_{0A}), and (3) How does z_0 vary with snow-covered area based on the underlying terrain?

2. Materials and Methods

The capability of a rough surface to absorb momentum from a turbulent boundary layer can be quantified by z_0, which is a measure of the vertical turbulence that occurs when a horizontal wind flows over a rough surface [32]. In general, z_0 is a quantity that is computed from the Reynolds number and the roughness geometry of the surface [29]. For rough, turbulent regimes occurring in the atmospheric boundary layer, dependence on the Reynolds number vanishes and z_0 is only a function of roughness geometry [33]. Various relations have been found to relate the geometry of roughness elements with z_0 [2,29]. For example, the dependence of z_0 on the size, shape, density, and distribution of surface elements has been studied using wind tunnels, analytical investigations, numerical modeling, and field observations [34,35]. Smith [36] provides a comprehensive review of the different approaches and models developed to analyze surface roughness and highlights that almost all models were developed for simplistic natural surfaces (i.e., regular arrays of roughness elements).The lack of a clear method for calculating z_0 as a function of surface roughness is due to the complexity of surfaces that exists in nature and the direction, spatial, and temporal dependence.

The most robust approach for estimating z_0 is from the anemometric method used to generate a logarithmic wind profile and solve for z_0 [32]. The anemometric method can be used for any surface with any arrangement of roughness elements but requires a meteorological tower of at least two vertically spaced wind, temperature, and humidity measurements that can be used to approximate the respective gradients. The measurements integrate over a footprint area rather than the single-point location of the sensors based on the distance from measurement source, elevation of sensor, meteorological conditions, turbulent boundary layer, and atmospheric stability. All of these factors can potentially create turbulent fluctuations affecting the downwind measurements of the wind profile [37,38]. The anemometric method is also very sensitive to the wind measurement heights; Munro [2] found that adding 0.1 m to any of the heights can alter z_0 by an order of magnitude.

In contrast, the geometric method uses algorithms relating z_0 to characteristics of surface roughness elements and thus does not require tower instrumentation but only a measure of the geometry of the surface [29].

Anemometric data are used to estimate z_0 from the logarithmic wind profile through an empirical relation that describes the vertical distribution of horizontal wind speeds within the lowest portion of the planetary boundary layer [39]. The wind speed (U_z in m/s) at height z (in m) above a surface is given by:

$$U_z = \frac{U^*}{k} \ln\left[\frac{z}{z_0} + \psi\left(z, \frac{z}{L}, L\right)\right] \tag{1}$$

where U^* is the friction velocity (m/s), k is the Von Kármán constant (~0.40), and ψ is a stability term, and L is the Monin-Obukhov stability parameter. This equation is only valid through the hypothesis of stationarity and horizontal homogeneity. Under neutral stability conditions, z/L tends towards zero, and ψ can be neglected.

The most common geometric method for estimating z_0 is simply a function of the height of the elements:

$$z_0 = f_0 z_h \tag{2}$$

where z_h is the mean height of roughness elements in meters, and f_0 is an empirical coefficient derived from observation [28]. The frontal area index, which combines mean height and breadth (all in meters), and density of the roughness elements, is defined as roughness area density given by:

$$(\lambda F) = Ly\, z_h\, \rho el \tag{3}$$

where Ly is the mean breadth of the roughness elements perpendicular to the wind direction, and ρel is the density or number (n) of roughness elements per unit area [40]. Lettau [29] developed a formula for z_0 based on the geometry of the surface for irregular arrays of reasonably homogenous elements:

$$z_0 = 0.5\, z_h\, \lambda\, F \tag{4}$$

In the Lettau formula, the coefficient 0.5 represents an average drag coefficient for the roughness elements, which was determined experimentally. Other geometric methods have been developed, especially to consider more regularly-shaped and distributed roughness elements, such as buildings in an urban setting [41,42]. The Counihan equation provides a geometric estimate of z_0 as:

$$z_0 = z_h\left(1.8\,\frac{A_f}{A_d} - 0.08\right) \tag{5}$$

where A_f is the total area in square meters silhouetted by the roughness elements, and A_d is the total area covered by roughness elements.

A meteorological tower was erected at Colorado State University Agricultural Research, Development and Education Center (ARDEC) South (http://aes-ardec.agsci.colostate.edu/), (40.629680, −104.99699) on a flat field that had no obstructions at least 100 m in the prevailing wind direction. The fetch was 40 m wide with the tower placed in the middle, leaving 100 unobstructed, homogenous meters upwind. Ten anemometers and five temperature and relative humidity sensors were placed vertically at different heights on the tower. The accuracy of the air temperature and relative humidity sensors (METER VP-3) was variable across a range of ±0.25–0.50 °C and ±4%, respectively (see http://manuals.decagon.com/Manuals/14053_VP-3_Web.pdf for more information). The METER Davis Cup Anemometers have a wind direction accuracy of ±7° and a speed accuracy within ±5% (see http://manuals.decagon.com/Manuals/). Data were collected from February 2014 through March 2015. In mid-March 2014, the flat field was plowed to create additional underlying roughness, specifically furrows and troughs were formed perpendicular to the dominant wind direction at an approximate spacing of two meters. The approximate amplitude of the troughs and furrows was 25 cm deep and 50 cm wide.

Meteorological data were recorded every five minutes based on the average of one-minute observations. Anemometric data were evaluated for 153 instances when wind speeds were faster than 4 m/s to ensure neutral stability [8] and when the log-linear fit had an r^2 greater than 0.95. The height of the instruments was calculated based on the depth of snow, which did not exceed 10 cm.

This study estimated z_0 values from anemometric measurements and used them as a reference to evaluate concurrent geometric methods. The z_{0A} values were calculated using Equation (1) from logarithmic anemometer wind profile data. Surface elevation was measured using a FARO Focus3D X 130 model TLS (https://www.faro.com/products/). This lidar tool generates a point cloud scan of a given area with an error of ±2 mm and a resolution of approximately 7.5 mm. The TLS was set up in 2–3 locations around the area of interest with 6 reference spheres to match the images using the FARO Scene Software. The data were generated into a point cloud and interpolated to a solid surface with 10 mm resolution with the kriging method using the Golden Software's Surfer 8 (https://www.goldensoftware.com/products/surfer). The gridded data were de-trended in the x-y plane to remove the bias in slope of the field or the angle of the lidar. Gaps in the scans tended to be small (<100 mm), and the kriging interpolation eliminated them. Individual roughness elements were identified and for each element the silhouette lot area and obstacle height were determined using a MATLAB code (https://www.mathworks.com/products/matlab.html). This was required to compute the Lettau formula (Equation (4)). The 1000-m^2 area around the tower was scanned on 12 occasions when the concurrent anemometric and geometric measurements were acquired. One concurrent measurement set was made with no snow cover for each of the unplowed and plowed scenarios; seven concurrent measurement sets were made with partial snow-covered area (SCA) and three with complete snow cover. SCA was determined from digital photos taken from the TLS unit.

Both the Counihan and Lettau methods were used to calculate z_{0G} (Equations (4) and (5), respectively). The Counihan method was appropriate for this study because the roughness elements (furrows) in this study site were semi-regular. During each concurrent anemometric and geometric measurement set, the percentage of the area covered in snow, or SCA, was estimated from photographs.

3. Results

The unplowed versus plowed field yielded different z_{0A} values (Figure 1). On average, the plowed field was almost 20 times as rough as the unplowed field, yet the coefficient of variation (COV) was essentially the same (0.67 and 0.62, respectively) (Figure 1). The smallest z_{0A} values for the plowed field were of the same magnitude as some of the largest z_{0A} values for the unplowed field, in the range of 1 to 3 $\times$ 10^{-3} m.

The Counihan method estimated z_{0G} values that were 1.39 times larger and had greater variation than the estimated z_{0A} values (Figure 2). We used the Nash-Sutcliffe coefficient of efficiency (NSCE), which is a performance statistic based on a comparison of the data fit to the 1:1 line, to evaluate how estimates of z_{0G} compared with z_{0A} [43]. The NSCE of the Counihan z_{0G} was -1.18, and the Lettau z_{0G} was 0.14, indicating the Lettau method compared more favorably with the z_{0A}. A linear regression between both z_{0G} estimates (Counihan z_{0G} and Lettau z_{0G}) and z_{0A} was fit through the data origin to evaluate if the bias between the two methods could be removed through simple linear scaling (Figure 2). When the Counihan z_{0G} values were scaled by 0.721 (1/1.39), the NSCE value only increased to 0.07. However, the NSCE increased to 0.88 when the Lettau z_{0G} values were scaled by 2.34 (1/0.428).

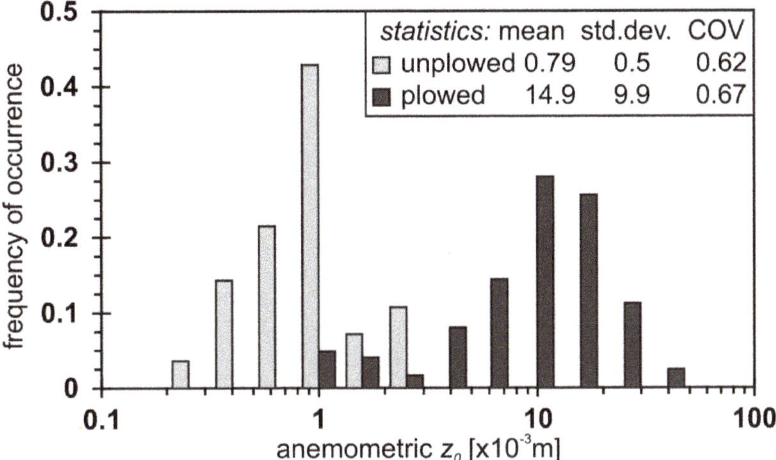

Figure 1. Histogram showing range and distribution of anemometric z_0 (z_{0A}) values for the unplowed and plowed field from anemometric data, based on 28 and 125 wind-speed profiles, respectively. The summary statistics (mean, standard deviation (std.dev.), and coefficient of variation (COV)) are presented in the legend. A logarithmic scale is shown on the x-axis to highlight the large difference for z_{0A} values among fields with varying characteristics.

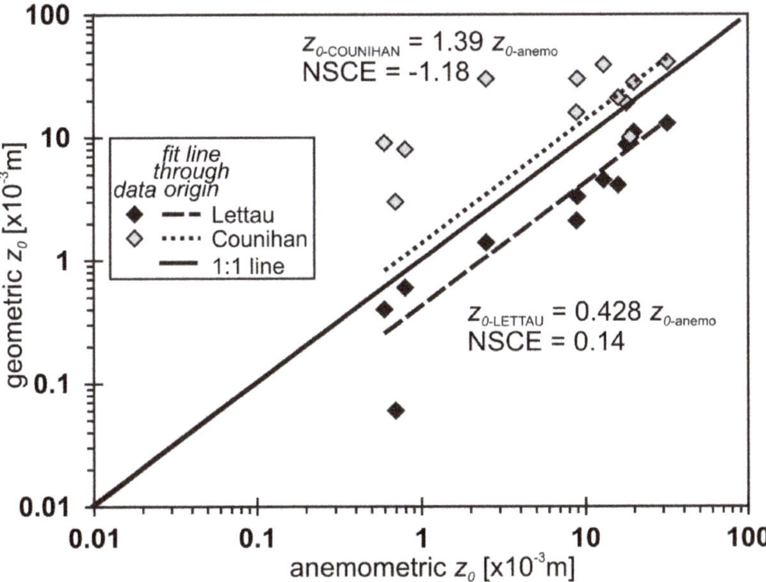

Figure 2. Comparison of Lettau and Counihan geometric methods to the anemometric method. A linear regression between the geometric-based Lettau and Counihan methods (z_{0G}) and the anemometric method (z_{0A}) fit through the origin are presented. The Nash-Sutcliffe coefficient of efficiency [43] fit statistic is also presented. When the Lettau method is scaled by 2.34 (1/0.428), the NSCE increases to 0.88. For the Counihan comparison, when it is scaled by 0.721 (1/1.39), NSCE increases to 0.07.

The estimated z_0 values were found to vary as a function of the amount of SCA present (Figure 3). As SCA increases, z_0 decreases, with variability based on the calculation method (Figure 3a). A linear regression between SCA and each of the z_0 estimates showed r^2 values that were 0.01, 0.7, and 0.88 for the Counihan, anemometric, and Lettau methods of z_0 calculation, respectively. There were noticeable differences in z_0 depending whether SCA was increasing because snow was accumulating versus when SCA was decreasing because the snow was melting. For periods of snow accumulation, removing snow measurements that were not immediately following a snow event (the yellow boxes in Figure 3b that represent non-accumulation values) improved the linear relation between accumulating SCA and z_0 ($R^2 = 0.94$).

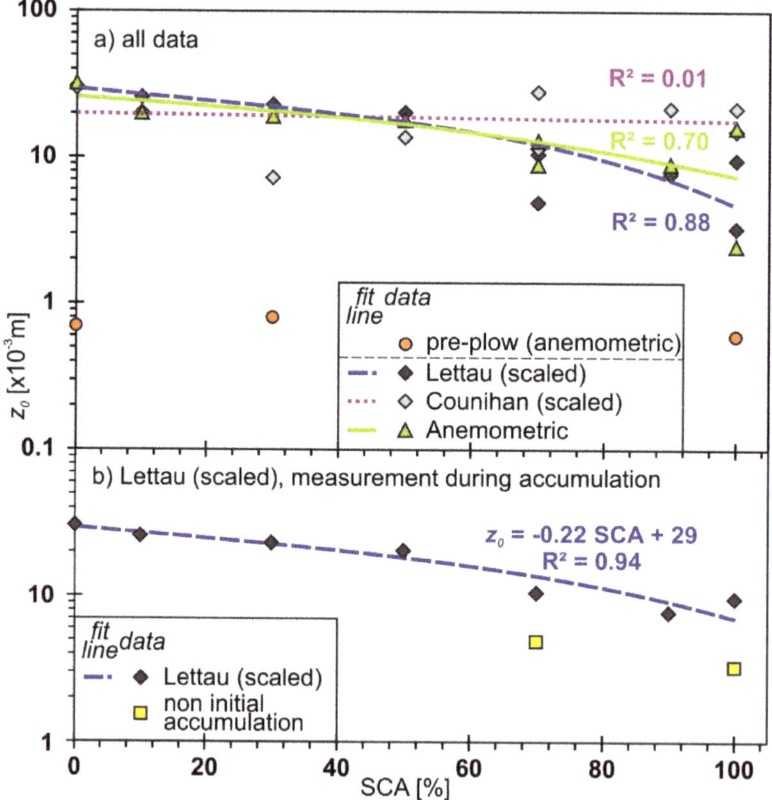

Figure 3. Linear relation between z_0 and snow-covered area (SCA as a %) for (**a**) all datasets (scaled Lettau and Counihan geometry-based and anemometric-based) with anemometric-based z_0 for the pre-plowed (orange circles) and plowed fields (green triangles) highlighted, and (**b**) the scaled Lettau geometry-based z_0 using a factor of 2.34 (see Figure 2). Lettau geometry-based z_0 measurements with non-accumulation snow measurements were removed. Lines are based on the best-fit linear regression of the data. Snow had been on the ground for numerous days prior to the two concurrent measurements (yellow squares) taken on 22 March 2014 (SCA = 100%) and 13 April 2014 (SCA = 70%). The snowfall was fresh for all other measurements.

4. Discussion

Geometrically estimated z_0, although easier to measure, produced different values when compared to the anemometric derived values. The Counihan method overestimated by a factor of 0.721, whereas the Lettau method underestimated anemometric z_0 (Figure 2) by a factor of 2.34. The Lettau method (Equation (4)) has a constant of 0.5 based on the average drag coefficient of the roughness characteristic of the silhouetted area of the average obstacle. By dividing the Lettau based

z_0 values by the 0.5 and thus eliminating the drag coefficient from the equation, we get a new NSCE of 0.856, with scatter in the data much closer to the 1:1 line (Figure 2). The removal of the drag coefficient suggests that the geometric data generated from the lidar point cloud appears to account for spatial and temporal variability in the roughness of a snow surface.

Lidar-based snow data are becoming more readily available [19,26]. The accuracy of the scans from about 1 mm for terrestrial lidar to 10 cm for airborne lidar can account for fine-scale changes of the snowpack [26], which enables the computation of z_0 at any scale. Although anemometric data can yield reliable estimates of z_0, meteorological towers are expensive to set up and operate. In addition, data from a single tower does not consider spatial variability as well as the geometric method [44]. Comparing the two methods does not consider the scale of the study area; the geometric measurement is taken over the entire area near the meteorological tower whereas the anemometric measurement is only influenced by the fetch area upwind of the sensors [29].

The roughness of the snowpack can vary substantially both spatially and temporally creating many implications [13,14,45]. Roughness variations can be caused by heterogeneities in land cover, vegetation, and meteorological conditions [46]; non-uniform distribution of snow cover during accumulation and melt [45,46]; snow-canopy interactions [47]; and snow redistribution by wind [48]. This was apparent in differences between the estimated z_0 for the plowed versus unplowed field (Figure 1). Land cover varies throughout regions particularly those with a shallow snow environment, and this creates variations that depend on the underlying topography [13,14,46]. Thus, there are many different values of z_0 in the literature [7] that are broader than our observed mean range of 0.2 to 10×10^{-3} m (Figure 1). For example, Miles et al. [31] found the z_0 of a hummocky glacier (a particularly rough underlying surface) to range between 5 to 500×10^{-3} m, whereas Brock et al. [7] reported z_0 values for fresh snow and older snow of 0.2×10^{-3} and 3.56×10^{-3} m, respectively. Our results show that change in roughness between a plowed and unplowed field yielded a 20-fold difference in z_0. The results of this study can be applied to areas of similar climate and land cover, which included flat, bare soil, and bare soil with small furrows (<1 m); and therefore, the results of this study may not scale appropriately to different land cover types. Further studies of a shallow snowpack in sagebrush steppe [49], farmland, or non-densely forested environments may be able to replicate our study results and scale from 1000 m^2 to a larger area. The z_0 values observed here had a notable change between flat soil and small furrows, so the changes in z_0 values in different environments with even minimal vegetation will have much larger effects on the z_0 values.

The inverse relation of SCA and z_0 (Figure 3) [50] is affected by the underlying terrain and size of the roughness features. As the snow accumulation increases, the roughness elements become buried, and the topography appears to be smooth [50,51]. This relation indicates that as snow accumulates over topographic features the snow will begin to level out at a z_0 height dependent scale. A hysteresis can be noted, and it has been found that a single snowfall event on a hummocky glacier can alter the micro-topography by up to 75% due to the shallow snowpack over the small scaled features [45,52]. The CLM4 uses a z_0 value of 2.4×10^{-3} m, a value that falls near the mean of the unplowed field, which is applicable for deep, flat snowpack surface with minimal influences from underlying terrain. However, this is not typical for shallower snowpacks or in complex terrains.

Relations between z_0 and SCA (Figure 3) can be used to improve snow-energy balance modeling by estimating the percentage of SCA via remote sensing and applying z_0 only to the portion of area it accurately describes [46,53]. Currently, most models use 100% SCA even though many areas will remain snow free due to complex terrain and can drastically change during periods of melt and accumulation [13,53]. Aerodynamic roughness length is incorporated into many climate and energy models, which require sub-grid snow distribution [54] and are still inadequate at representing SCA [46,48]. A dynamic z_0 based on SCA and land cover type can improve these on a sub-grid scale. Another complication with these models is the lack of accountability for snowpack variability throughout accumulation and melt [48,53,54].

Resolution is an important factor to consider when discussing both SCA and z_{0G}. The higher the resolution of the measurements (lidar, satellite, etc.), the higher the $z_{0\text{-}G}$ accuracy. However, lidar datasets are often large, especially those acquired with TLS, making them difficult and time consuming to process. Lower resolution data from remote sensing or airborne lidar systems (ALS) can cause problems when scaling [53]. Quincey et al. [52] found that z_{0G} is typically underestimated with a small area and coarse resolution and overestimated with a large area and fine resolution when compared to anemometric data. Nonetheless, even with lower resolution, applying dynamic z_0 values may greatly improve models. Scaling can be an effective way to incorporate both an anemometric and geometric z_0 value. Based on a specific land cover type, a scaling factor can be applied to areas with the same land cover. This can help to improve modeled z_0 accuracy, once preliminary z_0 values have been established.

5. Conclusions

Aerodynamic roughness length within our study has shown variation spatially and temporally for a shallow accumulating snow environment. This was apparent in our results that showed differences in z_0 mean values of about 14×10^{-3} m between the plowed and unplowed field. Thus, single-point measurements of anemometric data may not account for z_0 over a range of spatial and temporal scales. Geometrically calculated z_0 using the Lettau method has shown to be an effective and more robust form of z_0 estimation compared to the anemometric method and also producing similar, estimated values. The anemometric, single-point measurements will also not account for the snow-covered area, which changes based on its inverse relation with z_0. However, SCA can be observed and estimated from satellite imagery or airborne lidar systems to create a more accurate estimation of z_0.

Author Contributions: The experiments were designed by D.J.K., S.R.F., and W.L.B.; D.J.K. collected the field data; D.J.K. and S.R.F. did the analysis; W.L.B. helped with instrumentation setup; I.O. prepared the code to compute the geometric roughness length; J.E.S. and S.R.F. wrote the paper with input from G.A.S.; S.R.F. created the figures.

Funding: This research has been partially supported by the National Science Foundation under Grant No. DMS-1615909.

Acknowledgments: We thank the Colorado State University Water Center and the Colorado Water Institute for their financial support, especially for the installation of the anemometric tower. The Warner College of Natural Resources provided the Faro Terrestrial Lidar Scanner. Thanks are due to Dr. Edgar Andreas who provided some very insightful discussions about the movement of air over snow. We thank the referees for their useful comments that helped to improve the presentation of the paper. Partial funding for SRF was provided by the National Science Foundation project *"Pattern Formation and Spatiotemporal Complex Dynamics in Extended Anisotropic Systems"* (PI Iuliana Oprea; award number DMS-1615909). Any use of trade, firm, or product names is for descriptive purposes only and does not imply endorsement by the U.S. Government. All data are available at Colorado State University by request to the corresponding author.

Conflicts of Interest: The authors declare no conflict of interest.

References

1. Mialon, A.; Royer, A.; Fily, M. Wetland seasonal dynamics and inter-annual variability over northern high latitudes, derived from microwave satellite data. *J. Geophys. Res.* **2005**, *110*, 1–10. [CrossRef]
2. Munro, D.S. Surface roughness and bulk heat transfer on a glacier: Comparison with eddy correlation. *J. Glaciol.* **1989**, *35*, 343–348. [CrossRef]
3. Lehning, M.; Fierz, C. Assessment of snow transport in avalanche terrain. *Cold Reg. Sci. Technol.* **2008**, *51*, 240–252. [CrossRef]
4. Amory, C.; Naaim-Bouvet, F.; Gallee, H.; Vignon, E. Breif communication: Two well-marked cases of aerodynamic adjustment of sastrugi. *Cryosphere* **2016**, *10*, 743–750. [CrossRef]
5. Manes, C.; Guala, M.; Löwe, H.; Bartlett, S.; Egli, L.; Lehning, M. Statistical properties of fresh snow roughness. *Water Resour. Res.* **2008**, *44*, 1–9. [CrossRef]
6. Gromke, C.; Manes, C.; Walter, B.; Lehning, M.; Guala, M. Aerodynamic roughness length of fresh snow. *Bound.-Lay. Meteorol.* **2011**, *141*, 21–34. [CrossRef]

7. Brock, B.; Willis, I.; Sharp, M. Measurement and parameterization of aerodynamic roughness length variations at Haut Glacier d'Arolla, Switzerland. *J. Glaciol.* **2006**, *52*, 281–297. [CrossRef]
8. Andreas, E.L. Parameterizing scalar transfer over snow and ice: A review. *J. Hydrometeorol.* **2002**, *3*, 417–432. [CrossRef]
9. Fassnacht, S.R. Temporal changes in small scale snowpack surface roughness length for sublimation estimates in hydrological modeling. *J. Geogr. Res.* **2010**, *36*, 43–57. [CrossRef]
10. Rott, H. The analysis of backscattering properties from SAR data of mountain regions. *IEEE J. Ocean. Eng.* **1984**, *9*, 347–355. [CrossRef]
11. Williams, L.D.; Gallagher, J.G.; Sugden, D.E.; Birnie, R.V. Surface snow properties effect on millimeter-wave backscatter. *IEEE Geosci. Remote Sens. Soc.* **1988**, *26*, 300–306. [CrossRef]
12. Lacroix, P.; Legrésy, B.; Langley, K.; Hamran, S.E.; Kohler, J.; Roques, S.; Dechambre, M. In situ measurements of snow surface roughness using a laser profiler. *J. Glaciol.* **2008**, *54*, 753–762. [CrossRef]
13. Fassnacht, S.R.; Williams, M.W.; Corrao, M.V. Changes in the surface roughness of snow from millimetre to metre scales. *Ecol. Complex.* **2009**, *6*, 221–229. [CrossRef]
14. Fassnacht, S.R.; Stednick, J.D.; Deems, J.S.; Corrao, M.V. Metrics for assessing snow surface roughness from digital imagery. *Water Resour. Res.* **2009**, *45*. [CrossRef]
15. Fassnacht, S.R.; Toro Velasco, M.; Meiman, P.J.; Whitt, Z.C. The effect of aeolian deposition on the surface roughness of melting snow, Byers Peninsula, Antarctica. *Hydrol. Process.* **2010**, *24*, 2007–2013. [CrossRef]
16. Manninen, T.; Anttila, K.; Karjalainen, T.; Lahtinen, P. Instruments and methods automatic snow surface roughness estimation using digital photos. *J. Glaciol.* **2012**, *58*, 993–1007. [CrossRef]
17. Deems, J.S.; Fassnacht, S.R.; Elder, K.J. Fractal distribution of snow depth from lidar data. *J. Hydrometeorol.* **2006**, *7*, 285–297. [CrossRef]
18. Cline, D.; Yueh, S.; Chapman, B.; Stankov, B.; Gasiewski, A.; Masters, D.; Elder, K.; Kelly, R.; Painter, T.H.; Miller, S.; et al. NASA cold processes experiment (CLPX 2002/03): Airborne remote sensing. *J. Hydrometeorol.* **2009**, *10*, 338–346. [CrossRef]
19. Harpold, A.A.; Guo, Q.; Molotch, N.; Brooks, P.D.; Bales, R.; Fernandez-Diaz, J.C.; Musselman, K.N.; Swetnam, T.L.; Kirchner, P.; Meadows, M.W.; et al. LiDAR-derived snowpack data sets from mixed conifer forests across the Western United States. *Water Resour. Res.* **2014**, *50*, 2749–2755. [CrossRef]
20. Hood, J.L.; Hayashi, M. Assessing the application of a laser range finder for determining snow depth in inaccessible alpine terrain. *Hydrol. Earth Syst. Sci.* **2010**, *14*, 901. [CrossRef]
21. Prokop, A. Assessing the applicability of terrestrial laser scanning for spatial snow depth measurements. *Cold Reg. Sci. Technol.* **2008**, *54*, 155–163. [CrossRef]
22. Revuelto, J.; Lopez-Mureno, J.I.; Azorin-Molina, C.; Zabalza, J.; Arguedas, G.; Vicente-Serrano, S.M. Mapping the annual evolution of snow depth in a small catchment in the Pyrenees using the long-range terrestrial laser scanning. *J. Maps* **2014**, *10*, 379–393. [CrossRef]
23. López-Moreno, J.I.; Boike, J.; Sanchez-Lorenzo, A.; Pomeroy, J.W. Impact of climate warming on snow processes in Ny-Ålesund, a polar maritime site at Svalbard. *Glob. Planet. Chang.* **2016**, *146*, 10–21. [CrossRef]
24. López-Moreno, J.I.; Revuelto, J.; Alonso-Gonzalez, E.; Sanmiguel-Vallelado, A.; Fassnacht, S.R. Using very long-range terrestrial laser scanner to analyze the temporal consistency of the snowpack distribution in a high mountain environment. *J. Mt. Sci.* **2017**, *14*, 823–842. [CrossRef]
25. Nolan, M.; Larsen, C.F.; Strum, M. Mapping snow-depth from manned-aircraft on landscape scales at centimeter resolution using structure-from-motion photogrammetry. *Cryosphere Discuss.* **2015**, *9*, 333–381. [CrossRef]
26. Deems, J.S.; Painter, T.; Finnegan, D. Lidar measurement of snow depth: A review. *J. Glaciol.* **2013**, *59*, 467–479. [CrossRef]
27. Fassnacht, S.R.; Oprea, I.; Borleske, G.; Kamin, D.J. Comparing snow surface roughness metrics with a geometric-based roughness length. *Proc. Hydrol. Days* **2014**, *34*, 44–52.
28. Fassnacht, S.R.; Records, R.M. Large snowmelt versus rainfall events in the mountains. *J. Geophys. Res.* **2015**, *120*, 2375–2381. [CrossRef]
29. Lettau, H. Note on aerodynamic roughness-parameter estimation on the basis of roughness-element description. *J. Appl. Meteorol.* **1969**, *8*, 828–832. [CrossRef]
30. Andreas, E.L. A relationship between the aerodynamic and physical roughness of winter sea ice. *Q. J. R. Meteorol. Soc.* **2011**, *137*, 1581–1588. [CrossRef]

31. Miles, E.S.; Steiner, J.F.; Brun, F. Highly variable aerodynamic roughness length (z_0) for a hummocky debris-covered glacier. *J. Geophys. Res. Atmos.* **2017**, *122*, 8447–8466. [CrossRef]
32. Jacobson, M.Z. *Fundamentals of Atmospheric Modeling*, 2nd ed.; Cambridge University Press: Cambridge, UK, 2005; ISBN 978-0-521-54865-6.
33. Raupach, M.R.; Antonia, R.A.; Rajagopalan, S. Rough-wall turbulent boundary layers. *Appl. Mech. Rev.* **1991**, *44*, 1–25. [CrossRef]
34. Grimmond, C.S.B.; Oke, T.R. Aerodynamic properties of urban areas derived from analysis of surface form. *J. Appl. Meteorol.* **1999**, *38*, 1262–1292. [CrossRef]
35. Foken, T. *Micrometeorology*; Springer-Verlag: Berlin, Germany; New York, NY, USA, 2008; ISBN 978-3-540-74666-9.
36. Smith, M. Roughness in the earth sciences. *Earth-Sci. Rev.* **2014**, *136*, 202–225. [CrossRef]
37. Schuepp, P.H.; Leclerc, M.Y.; Macpherson, J.I.; Desjardins, R.L. Footprint prediction of scalar fluxes from analytical solutions of the diffusion equation. *Appl. Mech. Rev.* **1990**, *50*, 355–373. [CrossRef]
38. Sexstone, G.A.; Clow, D.; Stannard, D.; Fassnacht, S.R. Comparison of methods for quantifying surface sublimation over seasonally snow-covered terrain. *Hydrol. Process.* **2016**, *30*, 3373–3389. [CrossRef]
39. Oke, T.R. *Boundary Layer Climates*, 2nd ed.; Cambridge University Press: Cambridge, UK, 1987; ISBN 0-415-04319-0.
40. Raupach, M.R. Drag and drag partition on rough surfaces. *Bound.-Lay. Meteorol.* **1992**, *60*, 375–395. [CrossRef]
41. Counihan, J. Wind tunnel determination of the roughness length as a function of the fetch and the roughness density of three-dimensional roughness elements. *Atmos. Environ.* **1971**, *5*, 637–642. [CrossRef]
42. Macdonald, R.W.; Griffiths, R.F.; Hall, D.J. An improved method for the estimation of surface roughness of obstacle arrays. *Atmos. Environ.* **1998**, *32*, 1857–1864. [CrossRef]
43. Nash, J.E.; Sutcliffe, J.V. River flow forecasting through conceptual models part I—A discussion of principles. *J. Hydrol.* **1970**, *10*, 282–290. [CrossRef]
44. Luce, C.H.; Tarboton, D.G.; Cooley, K.R. Sub-grid parameterization of snow distribution for an energy and mass balance snow cover model. *Hydrol. Process.* **1999**, *13*, 1921–1933. [CrossRef]
45. Luce, C.H.; Tarboton, D.G. The application of depletion curves for parameterization of subgrid variability of snow. *Hydrol. Process.* **2004**, *18*, 1409–1422. [CrossRef]
46. Niu, G.; Yang, Z.L. An observation-based formulation of snow cover fraction and its evaluation over large North American river basins. *J. Geophys. Res.* **2007**, *112*, 1–14. [CrossRef]
47. Moeser, D.; Stahli, M.; Jonas, T. Improved snow interception modeling using canopy parameters derived from airborne lidar data. *Water Resour. Res.* **2015**, *51*, 5041–5059. [CrossRef]
48. Liston, G.E. Representing sub-grid snow cover heterogeneities in regional and global models. *J. Clim.* **2004**, *17*, 1381–1397. [CrossRef]
49. Tedesche, M.E.; Fassnacht, S.R.; Meiman, P.J. Scales of Snow Depth Variability in High Elevation Rangeland Sagebrush. *Front. Earth Sci.* **2017**, *11*, 469–481. [CrossRef]
50. Magand, C.; Ducharne, A.; Le Moine, N. Introducing hysteresis in snow depletion curves to improve the water budget of a land surface model in an alpine catchment. *J. Hydrometeorol.* **2014**, *15*, 631–649. [CrossRef]
51. Fassnacht, S.R.; Sexstone, G.A.; Kashipazha, A.H.; López-Moreno, J.I.; Jasinski, M.F.; Kampf, S.K.; Von Thaden, B.C. Deriving snow-cover depletion curves for different spatial scales from remote sensing and snow telemetry data. *Hydrol. Process.* **2016**, *30*, 1708–1717. [CrossRef]
52. Quincey, D.; Smith, M.; Rounce, D.; Ross, A.; King, O.; Watson, C. Evaluating morphological estimates of the aerodynamic roughness of debris covered glacier ice. *Earth Surf. Process. Landf.* **2017**, *42*, 2541–2553. [CrossRef]
53. DeBeer, C.M.; Pomeroy, J.W. Influence of snowpack and melt energy heterogeneity on snow cover depletion and snowmelt runoff simulation in a cold mountain environment. *J. Hydrol.* **2017**, *553*, 199–213. [CrossRef]
54. Fassnacht, S.R.; Yang, Z.-L.; Snelgrove, K.R.; Soulis, E.D.; Kouwen, N. Effects of averaging and separating soil moisture and temperature in the presence of snow cover in a SVAT and hydrological model. *J. Hydrometeorol.* **2006**, *7*, 298–304. [CrossRef]

MDPI

St. Alban-Anlage 66

4052 Basel

Switzerland

Tel. +41 61 683 77 34

Fax +41 61 302 89 18

www.mdpi.com

Geosciences Editorial Office

E-mail: geosciences@mdpi.com

www.mdpi.com/journal/geosciences